A Citizen's Guide to Artificial Intelligence

A Citizen's Guide to Artificial Intelligence

John Zerilli

with John Danaher, James Maclaurin, Colin Gavaghan,
Alistair Knott, Joy Liddicoat, and Merel Noorman

The MIT Press
Cambridge, Massachusetts
London, England

This book was set in Stone Serif and Stone Sans by Westchester Publishing Services. Printed and bound in the United States of America.

Library of Congress Cataloging-in-Publication Data

Names: Zerilli, John, author. | Danaher, John, author. | Maclaurin, James, author. | Gavaghan, Colin, author. | Knott, Alistair, 1967– author. | Liddicoat, Joy, author. | Noorman, Merel E., author.
Title: A citizen's guide to artificial intelligence / John Zerilli, with John Danaher, James Maclaurin, Colin Gavaghan, Alistair Knott, Joy Liddicoat, and Merel Noorman.
Description: Cambridge, Massachusetts : The MIT Press, [2020] | Includes bibliographical references and index.
Identifiers: LCCN 2020005708 | ISBN 9780262044813 (hardcover)
Subjects: LCSH: Artificial intelligence--Social aspects--Popular works.
Classification: LCC Q334.7 .Z47 2020 | DDC 006.3--dc23
LC record available at https://lccn.loc.gov/2020005708

Contents

Preface

The past decade has witnessed an unprecedented acceleration in both the sophistication and uptake of various algorithmic decision tools. From music and TV show recommendations, product advertising, and opinion polling to medical diagnostics, university admissions, job placement, and financial services, the range of the potential application of these technologies is truly vast. And while the business sector has wasted no time getting on board, governments too have been steadily integrating algorithmic decision support systems into their daily operations. Many police and law enforcement agencies around the world, for example, have co-opted deep learning tools in an effort to optimize efficiency and (so they say) reduce human bias. But while the roll-out continues to gather momentum and enthusiasts have welcomed the dawn of a new era, not everyone is convinced. Must those awaiting the outcome of a health insurance claim or defendants seeking bail or parole simply take it on faith that a machine knows best? Can a machine deciding such matters as the likelihood of a criminal reoffending really be accurate, free from bias, and transparent in its operations?

This book is the result of a joint effort in trying to make sense of the new algorithmic world order. It's structured around ten core themes. These canvass such questions as, what is artificial intelligence? Can an AI explain its decisions? Can it be held legally responsible? Does it have agency? What kind of control should humans retain over such systems, and does it depend on the kind of decision being made? Is the law of privacy in need of a fundamental rethink now that data sharing is easier and far more common than even ten years ago? How do we address the potential for manipulation through targeted political advertising? Is the use of decision tools by governments interestingly different from their use in industry? Do states have unique obligations to their citizens in this regard? How best to regulate behemoths like Facebook, Google, and Apple? Is regulation the answer? What kind of regulation?

The authorship of the book is a little unusual and perhaps merits a word or two of explanation. Put simply, although we wanted to have the best people write on the topics of their expertise, we didn't want the book to be a clunky miscellany of chapters and jarring styles, as is all too often the case with edited collections. Producing an authoritative work for the general public was certainly important to us, but we were adamant that it should have a single arc and speak with one voice. So, we needed someone to write a good chunk of the book—to set the tone, if you will—and someone who'd then be happy to take on what the others had written and mold it to that style. Ideally these roles would fall to the same person, and John Zerilli agreed to be that person, contributing just under half of the material and reshaping the rest to jibe with what he'd written.

The result, we trust, will prove useful to the bemused citizen.

Acknowledgments

John Zerilli, James Maclaurin, Colin Gavaghan, Alistair Knott, and Joy Liddicoat wish to acknowledge the generous assistance provided by the New Zealand Law Foundation to the *AI and Law in New Zealand* project.

John Zerilli also wishes to thank the Leverhulme Centre for the Future of Intelligence in the University of Cambridge for hosting him during parts of the book's composition. John is also grateful for the generous support of DeepMind Ethics and Society.

Parts of chapter 2 and the preface are reprinted with permission from Springer Nature, *Philosophy and Technology*, "Transparency in Algorithmic and Human Decision-Making: Is There a Double Standard?" by John Zerilli, Alistair Knott, James Maclaurin, and Colin Gavaghan, copyright 2018. Parts of Chapter 5 are reprinted from Springer Nature, *Minds and Machines*, "Algorithmic Decision-Making and the Control Problem" by the same authors, copyright 2019.

Acknowledgments

Prologue: What's All the Fuss About?

There's always something about astonishing technical feats that gives us pause. In a way they're both humbling *and* ennobling. They remind us how powerless we are without them and yet how powerful we must have been to produce them.

Sometimes what initially look like promising gadgets fizzle out and are soon forgotten (who uses a Blackberry anymore?). At other times the opposite seems to happen: an ingenious invention seems lackluster but later proves its mettle. In the twentieth century, neither the futurist H. G. Wells nor the British Royal Navy thought that submarines would amount to much.[1] Today, stealth submarines are an indispensable part of naval operations. Barely a century earlier, Charles Babbage's designs for an "Analytical Engine" were dismissed as crackpot fantasies, even though history would eventually, and spectacularly, vindicate his ambition and foresight. The Analytical Engine was essentially the world's first programmable general-purpose computer. Remarkably, Babbage's designs also anticipated the von Neumann architecture of every standard desktop in use today, complete with separate memory, central processor, looping, and conditional branching.

Every so often, however, an invention is unveiled to much fanfare, and *deservedly* so—when for once the hype is actually justified. Hype is the default setting in tech circles because new technology by its very nature tends to generate hype. The trick is to peruse the catalogue calmly and pick out only those items that justify their hype.

Of course, if we knew how to do this, we'd be in a better position to answer more profound questions about human history and destiny. For instance, at what moment in history is "a new reality" finally allowed to sink in? When does it become obvious that a Rubicon has been crossed and that things will never quite be the same again? To some extent your answer will depend on which theory of history you subscribe to. Is the history of the past hundred years, for example, a smooth, continuous narrative in which, at any point along the way, the next three to ten years might have been predicted in general outline at least? Or is the history of the last century a history of interruptions, of fits and starts, of winding roads and unforeseen eventualities? If history is smooth—perhaps not entirely predictable, but unsurprising let's say—it's going to be much harder to pinpoint significant moments because they're likely to fall beneath our radar. Maybe the election of a new leader looked so like every previous election that there was no way of telling at the time what a significant occasion it was. History ticks over more or less smoothly, on this theory, so truly epoch-making events are likely to be disguised in the ebb and flow of the familiar and mundane. On the other hand, if you subscribe to the "bumpy" theory of history, it should be easier to spot a watershed moment. On the bumpy theory, watershed moments are taking place all the time. History is forever taking unexpected courses and loves flank moves.

The truth is probably somewhere in between. History—infuriatingly? thankfully?—is both full of surprises *and* mundane. It has taken turns that nobody could have seen coming, and yet fortunes are still occasionally made on the stock market. Nothing that has happened in the past has ever quite been unthinkable. As the old saying goes, "there is nothing new under the sun." This means that, whether it's a relationship, the stock market, or the history of the entire planet that is concerned, no one is really in any better position than anyone else to state with confidence *this time things are different—from now on things are not going to be the same again*. Knowing what is momentous and what is banal, what will prevail and what will fade, is really anybody's guess. You might read Marx and Engel's *Communist Manifesto*, or Jules Verne's *Twenty Thousand Leagues Under the Sea*, or Alvin Toffler's *Future Shock* and marvel at the uncanny resemblance between what they "foresaw" and certain aspects of the modern world. But they were as often as not mistaken about the future too. And if anything, their mispredictions can be amusing. The 1982 film *Blade Runner* depicted a 2019 boasting flying cars, extraterrestrial colonization, and (… wait for it …) *desk fans*!

And yet such questions about the future exercise us whenever there looks to be a rupture in the otherwise smooth, predictable, orderly flow of time. Precisely because technology tends to generate hype, every fresh advance allures and beguiles anew, forcing us to contemplate its tantalizing possibilities. In effect, every major example of new technology insinuates a question—could *this* be the next big thing, the game-changer?

That's what we'd like to ponder in this book. Artificial intelligence, or AI for short, has generated a staggering amount of hype in the past seven years. Is it the game-changer it's been cracked up to be? In what ways is it changing the game? Is it changing the game in a *good* way? With AI these questions can seem especially difficult. On the one hand, as Jamie Susskind points out, "we still don't have robots we would trust to cut our hair."[2] On the other hand, Richard and Daniel Susskind describe a team of US surgeons who, while still in the United States, remotely excised the gall bladder of a woman in France![3]

No less important than these questions, however, are those that affect us as *citizens*. What do we, as citizens, need to know about this technology? How is it likely to affect us as customers, tenants, aspiring home-owners, students, educators, patients, clients, prison inmates, members of ethnic and sexual minorities, voters in liberal democracies?

Human vs. Machine

To start off, it's important to get a sense of what AI is, both as a field and as a kind of technology. Chapter 1 will tackle this in more detail, but it's helpful at the start to say a little about what AI hopes to achieve and how well it's going.

AI comes in many stripes. The kind that has generated most of the hype in recent years can be described in broad terms as "machine learning." One of the most prevalent uses of machine learning algorithms is in prediction. Whenever someone's future movements or behavior have to be estimated to any degree of precision—such as when a judge has to predict whether a convicted criminal will re-offend or a bank manager has to determine the likelihood that a loan applicant will repay a loan—there's a good chance a computer algorithm will be lurking somewhere nearby. Indeed, many coercive state powers in liberal democratic societies actually *require* an assessment of risk before those powers can be exercised.[4] This is true not just in criminal justice, but in areas like public health, education, and welfare. Now how best to go about this task? One way—and for a long time the *only* way—has been to rely on what's called "professional" or "clinical" judgment. This involves getting someone with lots of experience in an area (like a judge, psychologist, or surgeon) to venture their best bet on what's likely to happen in the future. (Will this criminal re-offend? What are the odds that this patient will relapse? And so on.) Professional judgment is basically trained gut instinct. But another way to approach this guessing game is to use a more formal, structured, and disciplined method, usually informed by some sort of statistical knowledge. A very simple statistical approach in the context of welfare decisions might just be to take a survey of previous recipients of unemployment benefits and ask them how long it took them before they found work. Decision makers could then use these survey results to hone their estimates of average welfare dependency and tailor their welfare decisions accordingly. But here's the rub. Although some people's intuitions are no doubt very good and highly attuned by years of clinical practice, research indicates that *on the whole* gut instinct comes a clear second to statistical (or statistically informed) methods of prediction.[5] Enter the new-fangled next-generation of machine learning algorithms. Mind you, this application of machine learning isn't exactly new—some of the predictive risk technology attracting media focus today was already being used in the late 1990s and in fact has precursors dating back decades and even centuries, as we'll discuss. But the speed and power of computing as well as the availability of data have increased enormously over the past twenty years (that's why it's called "big data"!).

There's a lot more that can be said about the relative performance of algorithms and statistics on the one hand and people and their experience-honed intuitions on the other. Suffice to say that sometimes, in specific cases, human intuition has proved to be better than algorithms.[6] Our point for now is that, on the whole, algorithms *can* make many routine predictive exercises more precise, less prone to idiosyncratic distortion—and therefore fairer—than unaided human intuition. That at any rate is the promise of AI and advanced machine learning.

Now let's consider another application of this same technology. Object classification is just what its name implies: assigning an instance of something to a particular class of things. This is what you do when you recognize a four-legged furry creature in the distance as either a cat or a dog; you are assigning the instance of the cat in front of you to the class CAT and not to the class DOG or the class CAR or the class HELICOPTER. This is something we humans do extremely quickly, effortlessly, and precisely, at least under normal viewing conditions. The other thing to say about how humans fare as object classifiers is that we tend to do so *gracefully*. A property of human object classification is what cognitive and computer scientists sometimes call "graceful degradation," meaning that when we make object classification errors, we tend to *near-miss* our target, and *almost-but not-quite* recognize the object. Even when the visibility is poor, for instance, it's easy to imagine ourselves mistaking a cow for a horse or a tractor for a buggy. But has any sane person ever mistaken a horse for an airplane or a buggy for a person? Unlikely. Our guesses are generally plausible.

What about machine learning object classifiers? Here there's a mixed bag of results. It's interesting to compare the way humans fail with the way machine learning systems fail. If we mistakenly classify a dog (say a husky) as a wolf—an easy mistake to make if you're none too familiar with huskies—what would have led you to that misclassification? Presumably you'd have focused your attention on things like the eyes, ears, and snout, and concluded, "Yep, that's a wolf." It might surprise you to learn that this eminently reasonable route to misclassification isn't the one an AI system would be obliged to take in reaching the same conclusion. An AI could just as easily focus on the shape of the *background* image—the residual image left after subtracting the husky's face from the photo. If there's a lot of snow in the background, an AI might conclude it's looking at a wolf. If there's no snow, it might decide otherwise.[7] You may think this is an odd way to go about classifying canines, but this odd approach is almost certainly the one an AI would take if most of the many thousands of images of wolves it was trained on contained snow in the background and if most of the images of huskies didn't.

If the same technology used for predicting whether someone is eligible for unemployment benefits or at risk of reoffending lies behind the classifier that concentrates on the absence or presence of snow to determine whether a canine is a husky or a wolf, we have a problem. Actually, there are several problems here.

One is what statisticians call "selection" or "sampling" bias. The classifier above had a bias in that the sample of images it was trained on had too many images of wolves in the snow. A better training set would've featured greater lupine diversity! For the same reason, a face recognition system trained on white men will have a hard time recognizing the faces of black or Asian women.[8] We'll come back to the subject of bias in chapter 3.

Another problem here is that it's often very hard to say why an AI "decides" the way it does. Figuring out that a classifier has fixed its attention on the snow around an object rather than the object itself can be a tricky business. We'll talk about algorithmic "transparency" and "explainability" in chapter 2.

Another issue is responsibility, and potentially legal *liability*, for harm. What if one of these classifiers were to be installed in an autonomous vehicle? If one day the vehicle ran into a child, thinking it was a tree, who'd be to blame? The vehicle's machine learning system may have learned to classify such objects autonomously, without being programmed *exactly* how to do it by anyone—not even its developers. Certainly the developers shouldn't get off scot-free. They would have curated the machine's training data, and (we hope!) made sure to include numerous examples of important categories like children and trees. To that extent the developers will have *steered* the machine learning system toward particular behavior. But should the developer *always* be responsible for misclassifications? Does there ever come a point when it makes sense to hold an *algorithm* responsible, or legally liable, for the harm it causes? We'll discuss these questions in chapter 4.

Then there's the issue of control. What happens when a time-pressed judge in a back-logged court system needs to get through a long list of bail applications, and a machine learning tool can save them a lot of time if its "objective" and statistically "precise" recommendations are simply accepted at face value? Is there a danger that a judge in this situation will start uncritically deferring to the device, ignoring their own reasonable misgivings about what the tool recommends? After all, even if these systems are frequently better than chance, and perhaps even better than humans, they are still far from perfect. And when they do make mistakes—as we saw they can—they often make them in strange ways and for odd reasons. We'll discuss the control problem in chapter 5.

Another (huge!) issue is data privacy. Where did all that training data come from? Whose data was it to begin with, and did they consent to their data being used for training private algorithms? If anything, the 2020 COVID-19 global pandemic raises the stakes on such questions considerably. As this book goes to press, governments the world over are contemplating (or implementing) various biosurveillance measures that would enable them to track people's movements and trace their contacts using mobile phone data. This is all very well for crisis management, but what about when the crisis is over? Experience shows that security and surveillance measures can be scaled-up fairly quickly and efficiently if needed (as they were after the 9/11 attacks). But as the phrase "surveillance creep" suggests, their reversal is not approached with anything like the same alacrity. Chapter 6 delves into these and similar sorts of data protection issues.

A final question, and perhaps the most important one of all, relates to the effects that the use of such systems will have on human autonomy and agency in the longer term. We'll investigate these effects in chapter 7, but it's worth saying just a little about this topic now because there's a lot to ponder and it'll be useful clearing up a few things from the get-go.

One of the topics that's exercised AI folk in recent years has been the effect of increasingly sophisticated AI on human dignity. The phrase "human dignity" isn't easy to define, but it seems to be referring to the *worth* or *value* of human life. The question is whether advanced AI systems diminish the worth of human life in some way. The possibility that machines could one day reproduce and even exceed the most distinctive products of human ingenuity inspires the thought that human life is no longer special. What are we to make of such concerns?

As the object classifier example illustrates, machine learning tools, even quite sophisticated ones, aren't "thinking" in anything like the way humans think. So if object classifiers are anything to go by, it doesn't look like machines will compete with us on the basis of *how* we do things, even if they can do *what* we do in different ways. For all we know, this state of affairs might change in the future. But judging from today's technological vantage point, that future seems a long way off indeed. This ought to offer some reassurance to those concerned for human dignity.

But is that enough? You might think that if a machine could come to do the kinds of things we pride ourselves on being able to do as human beings, what does it matter if a machine can perform these very same feats differently? After all, it's not *how* a mousetrap works but *that* it works that's important, isn't it? Maybe airplanes don't fly as elegantly or gracefully as eagles—but then why should that matter? Would an aircraft need to fly so like a pigeon that it would fool other pigeons into thinking it was a pigeon before we could say it could "fly"?[9] Hardly. So, to repeat, if AI can match or surpass the proudest achievements of humankind, albeit through alternative means (the way planes can rival avian flight without making use of feathers), is human dignity any the worse for that?

We don't think so. Human calculating abilities have been miles behind humble desktop calculators for many decades, and yet no one would seriously question the value of human life as a result. Even if machine learning systems start classifying objects more reliably than humans (and perhaps from vast distances, too), why should this diminish the worth of a human life? Airplanes fly, and we don't think any less of birds.

The philosopher Luciano Floridi helpfully reminds us of several other developments in history that threatened to diminish human dignity, or so people feared. He mentions how Nicolaus Copernicus, Charles Darwin, Sigmund Freud, and Alan Turing each in their own way dethroned humanity from an imagined universal centrality, and thus destabilized the prevailing and long-held "anthropocentric" view of nature.[10] Copernicus showed that we are not at the center of the universe; Darwin showed that we are not at the center of the biosphere; Freud showed that we are not at the center of the psychosphere or "space of reason" (i.e., we can act from unknown and introspectively opaque motivations); and Turing showed that we are not at the center of the infosphere. These cumulative blows weren't easy to take, and the first two in particular were violently resisted (indeed the second still so). But what these revelations did *not* do—for all the tremors they sent out—was demonstrate the idea of human dignity itself to be incoherent. Of course, if human dignity means "central," you certainly could say human dignity was imperiled by these events. But dignity and centrality aren't the same. An object doesn't have to be at the center of a painting for it to capture our attention or elicit admiration.

There's another reason why, for the foreseeable future, human dignity is likely to be unaffected by the mere fact that machines can beat us at our own game, so to speak. Every major AI in existence today is *domain-specific*. Whether it's a system for beating chess champions or coordinating transactions on the stock exchange, every stupendous achievement that has been celebrated in recent years has occurred within a very narrow domain of endeavor (chess moves, stock trades, and so on). Human intelligence isn't like this—it's *domain-general*. Our ability to play chess doesn't preclude our ability to play tennis, or write a song, or bake a pie. Most of us can do any of these things if we want to. Indeed many researchers would regard the holy grail of AI research as being able to crack this domain-general code. What is it about human minds and bodies that make them able to adapt so well, so fluidly, to such divergent task demands? Computers do seem to find easy what we find hard (try calculating 5,749,987 × 9,220,866 in a hurry). But they also seem to experience staggering difficulty with what most of us find really easy (opening a door handle, pouring cereal from a box, etc.). Even being able to hold a conversation that isn't completely stilted and stereotyped is very difficult for a machine. When we ask our roommate to pick up some milk on the way home, we know this means stop by the corner store, purchase some milk, and bring it home. A computer needs to be programmed so as *not* to interpret this instruction in any number of far more literal—and so algorithmically simple—ways, such as find milk somewhere between your present location and home; upon finding it, elevate its position relative to the ground; then restore the milk to its original position. The first (and obviously sane) interpretation requires a subtle integration of linguistic and contextual cues that we manage effortlessly most of the time (linguists call this aspect of communication "pragmatics"). This simply isn't the case for machines. Computers can do syntax well, but the pragmatic aspects of communication are still mostly beyond them (though pragmatics *is* an active area of research among computational linguists).

The same can be said of consciousness. No existing AI is presently conscious, and no one's got any real clue just how or why conscious experiences arise from vacant material processes. How do you get the felt sense of an inner life, of an internal "point of view," from a group of proteins, salt, and water? Why does anything touchy-feely have to accompany matter at all? Why can't we just be dead on the inside—like zombies? There are many theories, but the precise character of consciousness remains elusive. For all the science-fiction films that play on our worst fears of robots becoming conscious, nobody's any closer to achieving Nathan Bateman's success with "Ava" in the 2015 film *Ex Machina*.

Thus for so long as the advent of sophisticated AI is limited to domain-specific and unconscious systems, we needn't be too worried about human dignity. This isn't to say that domain-specific systems are harmless and pose no threat to human dignity for *other* reasons—reasons unconnected with an AI's ability to trounce us in a game of chess. Lethal autonomous weapons systems are obviously attended by the gravest of dangers. New data collection and surveillance software also poses significant challenges to privacy and human rights. Nor are we saying that domain-specific technologies cannot lead to revisions in how humans perceive themselves. Clearly, the very existence of a system like Google DeepMind's AlphaGo—which thrashed the human Go world-champion in a clean sweep—must change the way humans think of themselves on some level. As we noted at the start of this prologue, these systems can reasonably evoke both pride and humility. Our only point here is that human dignity isn't likely to be compromised by the mere fact that our achievements can be matched or outdone.

Is AI Rational?

There are several other issues around AI we haven't mentioned and won't be exploring in this book. We've picked out the ones we thought would be of most interest. One other issue we'd like to briefly mention concerns what we might call the *rationality* of machine learning. The fact is, many machine learning techniques aim at the discovery of hidden relationships in data that are generally too difficult for humans to detect by themselves. That is, their whole modus operandi is to look for *correlations* in the data. Correlations are a legitimate source of knowledge, to be sure, but as most of us learn in high school: *correlation is not causation.* Two clocks might always strike twelve at the same time each day, but in no sense does one clock's striking cause the other clock's striking. Still, as long as a correlation is reliable—the technical term is "statistically significant"—it can provide actionable insights into the world despite not being causal. For example, it might be discovered that people who prioritize resistance training tend to make particular nutrition choices, eating on the whole more meat and dairy than those who prioritize aerobic forms of exercise. A fitness club could reasonably go off such insights when deciding what sorts of recipes to include in its monthly health magazine—even if, strictly speaking, no causal link between the type of exercise people do and the type of food they eat can be firmly established. Perhaps weight-trainers would personally prefer plant-based proteins, but as a demographic, it's simply easier for them to source animal-based proteins.

But at this point you might wonder: What if the correlations are reliable but also utterly bizarre? What if an algorithm discovers that people with small shoe sizes eat a particular type of breakfast cereal or that people with a certain hair color are more prone to violence than others? Such relationships could, of course, be merely incidental—flukes arising from poor quality or insufficiently large training data. But let's put that concern to one side. It's true that a machine learning system that finds bizarre correlations is likely to have been mistrained in ways familiar to statisticians and data scientists. But the whole point of a "statistically significant" correlation is that it accounts for—and, in theory, rules out—this possibility.

These aren't altogether fanciful illustrations, mind you. One of the benefits of unsupervised machine learning (see chapter 1) is that it can detect the sorts of correlations that we ourselves would never think to unearth. But in English law, if a public official were to decide a case on the basis of correlations as apparently spurious as these, the decision would be quashed. In a passage well-known to English and Commonwealth lawyers, Lord Greene once said that an official

> must exclude from his consideration matters which are irrelevant to what he has to consider. ... Similarly there may be something so absurd that no sensible person could ever dream that it lay within the powers of the authority. Warrington LJ ... gave the example of the red-haired teacher, dismissed because she had red hair. ... It is so unreasonable that it might almost be described as being done in bad faith.[11]

But, we suspect, *just these sorts of correlations* are likely to proliferate with the steady march of machine learning in all areas of public and private decision making: correlations that would seem untenable, illogical, and even drawn in bad faith if a human had been behind them. To be clear, we're not saying that such correlations will defy all attempts at explanation—we're willing to bet that a proper explanation of them will, in many cases, be forthcoming. But what are we to do in the meantime, or if they turn out *not* to be intuitively explicable after all? Let's imagine that an algorithm discovers that people who like fennel are more likely to default on their loans. Would we be justified in withholding credit from people who like fennel? Last time we checked, "liking fennel" wasn't a protected attribute under antidiscrimination law, but *should* it be? Denying someone a loan because they reveal a penchant for fennel certainly *looks* like a kind of discrimination: "liking fennel" could well be a genetically determined trait, and the trait certainly *seems* irrelevant to debt recovery. But what if it's not irrelevant? What if there's actually something to it?

In these and other ways, machine learning is posing fresh challenges to settled ways of thinking.

Citizen vs. Power

So if the biggest challenge posed by AI this century isn't the rise of a conscious race of robots, the brief for us as authors is to produce a book whose sweep is necessarily political rather than technical. We'll cover only so much of the technical background as is required to make the political issues come to life (in chapter 1).

A book with a political bent, of course, can't claim to be very useful if it doesn't offer at least a few suggestions about what might be done about the problems it diagnoses. Our suggestions will be liberally sprinkled throughout, but it's not until we hit chapter 10 that we'll discuss regulatory possibilities in more detail. As long ago as 1970, Alvin Toffler, the American futurist, wrote of the need for a technology ombudsman, "a public agency charged with receiving, investigating, and acting on complaints having to do with the irresponsible application of technology."[12] In effect he was calling for "a machinery for screening machines."[13] Well, that was 1970, a time when only the faintest rumblings of disruption could be heard and then only by the most acute and perceptive observer (as Toffler himself was). Today, the need for regulatory responses in some form or another—indeed probably taking a variety of forms—is not just urgent; it's almost too late. The fractiousness and toxicity of public discourse, incubated in an unregulated social media environment, have already contributed to the malaise of our times—reactionary politics, the amplification of disreputable and sectarian voices, and atavistic nihilism. So today we need to think much more boldly than we might have done in 1970. Indeed, the COVID-19 pandemic has made one thing brutally apparent: some challenges may require interventions so drastic that they could be seen to herald a new understanding of the relationship between a state and its citizens.

1 What Is Artificial Intelligence?

In today's terms, artificial intelligence (AI) is best understood as a special branch of computer science. Historically, the boundary between AI and computer science was in fact less clear-cut. The history of computer science is intimately linked with the history of AI. When Alan Turing invented the general-purpose computer in 1936 (on paper!), he was evidently thinking about how to model a person doing something clever—namely, a specific kind of mathematical calculation. Turing was also one of the first to speculate that the machine he invented for that purpose could be used to reproduce humanlike intelligence more generally. In this sense, the father of computer science is also the father of AI. But other computer pioneers also thought about computing machinery in human (indeed anthropomorphic) terms. For instance, John von Neumann, in a famous 1945 paper introducing the architecture still used by computers today, used a term from psychology, "memory," to refer to a computer's storage units.

In a sense, we can still think of computers as exhibiting "intelligence" in their general operation. By chaining together simple logical operations in myriad ways, they accomplish an extraordinary and impressive range of tasks. But these days, AI is very clearly a subfield of computer science. As AI and computer science evolved, AI focused more squarely on the tasks that humans are good at. A famous definition of AI states that AI is the science of making computers produce behaviors that would be considered intelligent *if done by humans*.[1] Of course, this is no easy undertaking. Computers, we noted in the prologue, are exceptionally good at many tasks we find extremely difficult, such as solving complex mathematical equations in split seconds, but often spectacularly bad at tasks we find a breeze, such as walking, opening doors, engaging in conversation, and the like.

This human-centered definition of AI still works pretty well, but needs a little updating. It's true that many of the central applications of modern AI involve reproducing human abilities, in language, perception, reasoning, and motor control. To this extent, AI involves making computers "in our own image," so to speak. But nowadays AI systems are also used in many other more arcane areas to accomplish tasks whose scale or speed far exceed human capabilities. For instance, they're used in high-frequency stock trading, internet search engines, and the operation of social media sites. In fact, it's useful to think of modern industrial-scale AI systems as possessing a mixture of both subhuman and superhuman abilities.

There's obviously a great deal more to AI than this potted history can get across, but it's enough to get you started and enough also to situate the focus of this book. In this book, we're going to focus on the most influential approach to AI today called "machine learning." We describe machine learning as an "approach" to AI because it's not concerned with any specific task or application. Rather, machine learning comprises a *set of techniques* for solving any number of AI problems. For instance, natural language processing, speech recognition, and computer vision have each been pursued within the general approach to AI known as machine learning.

As the name suggests, machine learning harnesses both the power of computers as well as the sheer volume of data now available (thanks to the internet) to enable computers to *learn for themselves*. In the usual case, what are learned are patterns in the data. From now on, when we refer to AI, we'll refer to machine learning, unless the context dictates otherwise.

The rest of this chapter will introduce and then deploy a few concepts that you might not have encountered since high school algebra. Some of you might not have encountered them at all. Here we have in mind concepts like a mathematical "function," a "parameter," and a "variable." These terms are sometimes used in ordinary speech, but they have specific, technical meanings in mathematics and computer science, and their technical definitions are what we'll mean when we use them. We'll do our best to explain the gist of these as we go along, as the gist is all you'll need. But if you find the explanations glib or unsatisfactory in any way, do take heart. We don't go into much detail precisely because all you need is the gist. So, if you're turned off by anything remotely mathy, get out of this chapter what you can, and then definitely proceed with the rest of the book. If you don't completely grasp something, chances are it won't matter. Again, the book is intended to be primarily political, not technical.

By the way, you can dip into this book at any point—each chapter can be read on its own without having read any of the others first. If there are relevant bits of other chapters we think you might like to read in conjunction, we'll refer back (or forward) to that chapter.

Machine Learning and Prediction

In this book, we're going to focus on a particular class of machine learning technologies, which we term "predictive models." This focus doesn't include all machine learning models, but it does encompass the technologies at the center of the current "AI revolution" that's causing so much disruption in industry and government. Importantly, it also encompasses systems that have been in use in industry and government for decades—and in some cases, much longer. In the media, AI is often portrayed as a brand new arrival—something that's suddenly, and recently, started to affect public life. In our introduction to AI, we want to emphasize that the AI models currently affecting public life are very much a *continuous* development of statistical methods that have been in use for a long time in both the private and public sectors. There is a long tradition of statistical modeling in both government and commerce, and the statistical models used historically in these fields actually have much in common with the predictive AI models that are now becoming prevalent. To emphasize these commonalities, it's helpful to focus on the predictive models that are at the core of modern AI.

Our focus on predictive models will also be helpful in emphasizing that the AI models used in government departments are technically very similar to those used in modern commercial settings. For instance, the models used in finance ministries to hunt for tax evaders or in justice departments to assess defendants' risks of reoffending are the same *kind* of model that Amazon uses to recommend books we might be interested in and that Google uses to decide which ads we see. When, toward the end of the book, we come to consider how AI techniques in government and business should be regulated, it will be helpful to understand that the technical methods under scrutiny in these two areas are largely the same. Regulation of the use of AI models in government may look quite different from regulation of their use in industry owing to the very different social functions of the institutions in these two spheres. But the models that are the *object* of regulation in these two areas are basically the same.

Predictive Modeling Basics

We'll begin with a simple definition of a predictive model. A predictive model is a mathematical procedure that makes predictions about the value of some *unknown variable* based on one or more *known variables*. A "variable" can be any measurable aspect of the world. For instance, a predictive model might predict a person's weight based on their height. Such a model would be useful if for some reason we can readily get information about people's height but not about their weight (and we are nevertheless *interested* in their weight).

Note that a predictive model doesn't have to predict an occurrence in the future. The unknown variable might relate to the current moment or even to times in the past. The key thing is that we need to *guess* it because we can't measure it directly (the word "guess" is probably more to the point than the word "predict" here). Thus we can define a predictive model as a tool that can make guesses about some *outcome variable* based on a set of *input variables*.

To build a predictive model, the key ingredient is a *training set* of cases where we know the outcome variables as well as the input variables. In the above example, the training set would be measurements of the height and weight of a number of sample people. There is a (loose) relationship between people's height and weight. The training set provides information about this relationship. A predictive model uses this information to compute a general hypothesis about the relationship between height and weight that it can use to make a guess about someone's weight given their height. A training set is in essence a database of facts about known cases. The larger this database, the more information is provided about the outcome variable. We will take it as part of the definition of a "predictive model" that it is derived from training data through a training process of some kind.

A Brief History of Predictive Models

Mathematical methods have been in use for centuries to guess an unknown variable by consulting a database of known facts. The first serious applications were in the insurance industry. The Lloyd's register, which developed in 1688 and assessed the likely risks of shipping ventures, is a well-known early example.[2] Our survey of predictive models will begin slightly later with the Equitable Life insurance company, which was the first company to use data systematically to make predictions. Equitable Life was founded in 1762.[3]

The earliest predictive models with relevance to government date from around this same time. For instance, in the 1740s the German statistician Johann Süssmilch used data from church records to devise a model that systematically used the availability of land in a given region to predict marriage age and marriage rate (and, through these, birth rates).[4] Governments have been maintaining databases of information about their citizens from time immemorial, largely for the purposes of assessing their tax obligations. As the science of predictive modeling developed, these databases could be reused for other government functions, particularly those relating to financial planning. The British Government employed its first official actuary in the 1830s, an employee who worked in naval pensions and is credited with saving the government hundreds of millions of pounds in today's money.[5] Already at that time, the predictive models used in government mirrored those used in business—a trend that continues to this day.

To begin with, developing predictive models involved calculations done by hand, using databases stored in written ledgers. Computers can help in two ways: they facilitate the storage of large amounts of data, and they can perform calculations automatically. Predictive models are now routinely implemented as computer programs that consult databases held in computer memory.

When computers were first introduced, their only users were governments and large corporations owing to their great expense. Both companies and governments started to develop computer-based predictive models almost as soon as computers were invented. For instance, the US government used computers to predict missile trajectories in the 1940s,[6] to predict weather in the 1950s,[7] and to predict suitability of military personnel for missions in the 1960s.[8] In industry, the FICO corporation in the United States, which specializes in predicting credit risk, produced its first computerized model of risk scores in 1958.

We will introduce modern predictive AI systems by first presenting some older predictive statistical models, and then showing how the AI models extend (and differ from) these.

Actuarial Tables

The Equitable Life company made a novel contribution to insurance when it produced a table showing, for each age, the probability that a person will die at that age (based on available mortality statistics), and computing an associated insurance premium for that age. This kind of table came to be known as an "actuarial table." This venerable insurance practice counts as a predictive model by our definition. The input variable is "age"; the outcome variable is "chance of dying"; and the training data are the mortality statistics used to compile the table. The innovation of the actuarial table lay in its *systematically* charting the outcome variable for each possible input variable within a given range.

As actuarial science progressed, more complex tables were developed, taking into account factors other than age so that more accurate predictions could be made and more accurate premiums charged. These tables implemented more complex predictive models with more input variables.

Geometric Approaches to Predictive Modeling

A drawback with actuarial tables is that they treat ages as discrete categories and then compute probabilities for each age separately, without regard for one another. But age varies continuously, and the probability of dying varies smoothly as a *function* of age. A function is a special type of relationship between input and output variables. What makes the relationship special is that the value of the output variable *depends on* and can be *determined by* the value of the input variable. In the earlier example we gave, height and weight are related in this special way. The "value" of your weight—that is, how much you weigh—is (partly) dependent on and can be determined by your height. The taller you are, the more you'll weigh. Many variables of interest to scientists in the real world are related in this way. In technical terms, we say that the relationship between such variables *defines* a function. And as you may recall from high school, these relationships can be plotted as curves on graphs (the famous Cartesian number plane). That is, we can think of functions *geometrically* as curves with different shapes.

Actually it's useful to be able to think about *probabilities* geometrically. For instance, we can define a mathematical function that maps the variable "age" represented on the *x* axis of a graph onto a probability of dying represented on the *y* axis, as shown in figure 1.1. This idea was pioneered by the nineteenth century actuary Benjamin Gompertz who found that the probability of dying can be modeled quite accurately by a simple mathematical function. Importantly, the function is *continuous*, meaning there are no breaks in it. For absolutely any age in figure 1.1, including, theoretically, something very precise like 7.78955 years of age, we can find the corresponding probability of dying. You wouldn't easily be able to do this using an actuarial table because the age increments would soon get unwieldy if years were expressed correct to three or four decimal places.

Note there are several curves in figure 1.1. Like many mathematical functions, Gompertz's function has various *parameters* that can be adjusted. The four curves in this figure show the same function with different sets of parameter values. You might remember from high school that the same function $y = 2x^2$ would come out looking different on a graph if, instead of a 2 in front of the x^2, you had a 4 in front of it. Here, 2 and 4 are called "parameters," but the underlying function $y = ax^2$ is the same function, regardless of what its parameters might be (i.e., regardless of the value of a). In this new geometric approach to prediction, we don't use our training data to estimate many separate probabilities, as we would in creating actuarial tables. Instead, we use the data to find the parameters of a function that best "fit" the data. If we model mortality with a "parameterizable" function, we can use our training set to find the parameter values that make the function most closely approximate the known facts about mortality. Of course, we have to begin by choosing a parameterizable function that's suitable for modeling our data

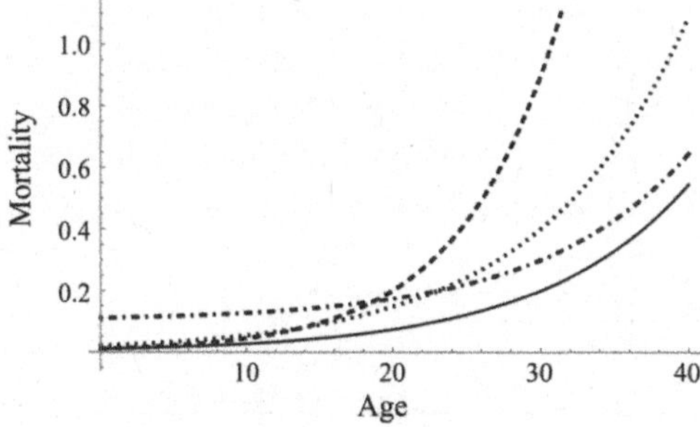

Figure 1.1
Examples of the Gompertz curve, a mathematical function mapping age onto probability of dying.

in the first place. But having made this choice, the process of finding the parameters that best fit the data can be automated in various ways. In this paradigm, the predictive model is nothing more than a mathematical function, with certain specified parameter values.

A particularly influential method for fitting a function to a given set of training data is called *regression*, which was first developed in the nineteenth century.

Regression Models

Take that "loose" relationship between height and weight we mentioned earlier. To quantify this relationship, we can gather a set of known height-weight pairs to serve as a "training set" for a model that maps height onto weight. An example training set is shown in figure 1.2, as a set of data points on a two-dimensional graph, where the *x* axis depicts height, and the *y* axis depicts weight.

Crucially, having represented the training set as points in a graph, we can "learn" a mathematical function that maps *every possible height* onto a weight, as shown by the red line in figure 1.2. (The training points include "noise,"

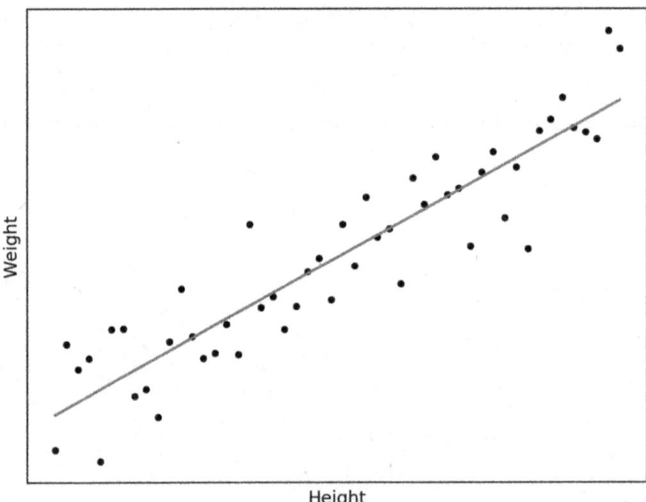

Figure 1.2
A function defining a relation between people's height and weight, learned by linear regression from the data points shown in black.

miscellaneous factors that are relevant to weight but aren't included in the model.) The line gives an answer even for heights that aren't included in the training set and thus can be used to estimate weights for people who aren't exactly like those in the training set. Note that the line doesn't go through many of the training points. (It might not go through any at all!) In the presence of noise, we have to make a "best guess."

Let's say we decide we're going to use a straight line to model our data. We can draw all sorts of straight lines on the graph in figure 1.2, but which of these most accurately represents the data points? Answering this amounts to finding the parameters of our straight-line function that best "fits" the training points. Linear regression is a way of finding the parameter values that provably minimize the amount of "error" the line makes in its approximation of the training points. The regression technique takes a set of training points and gives us the best parameter values for our function. We can then use this function as a general predictive model of the kind that is our focus—it will make a prediction about a weight for any given height.

Modern Regression Models

Regression is a key technique in modern statistical modeling. The basic method outlined above has been expanded in a multitude of different ways. For instance, linear regression models can involve many variables, not just two. If we have exactly three variables, for example, data points can be visualized in a three-dimensional space, and regression can be understood as identifying a three-dimensional *plane* that best fits the points. Moreover, regression modelers are free to decide how *complex* the function that fits the training data points should be. In figure 1.2, the function is a straight line, but we can also allow the function to be a curve, with different amounts of "squiggliness"—or, in three dimensions or more, a plane with different amounts of "hilliness."

Regression techniques can also be used to model relationships between variables that vary *discretely* rather than continuously. (An example of a discrete variable is the outcome of a coin toss: there are just two possibilities here, unlike the continuous range of possibilities for height or weight.) This is done using "logistic regression" models, which are useful for classification tasks (we'll say more about classification below when we discuss decision trees). Finally, there are many varieties of regression model specialized for particular tasks. An important variety for many government applications is "survival analysis," which is a method for estimating the likely amount of time that will elapse before some event of interest happens. The original applications of these methods were in drug trials in which the "event of interest" was the death of a patient under some form of therapy. But there are many applications in government or commercial planning in which it is very useful to have a way of predicting how far in the future some event may be for different people or groups. Again, the same regression methods are used both within government and industry.

We should also note that regression models don't *have* to be used in these action-guiding, practical ways. Scientists who use it are often just interested in stating relationships between variables in some domain. A scientist might, for instance, want to be able to state as an empirical finding that "there is a relationship between height and weight." Methods relating to regression can be used to quantify the strength of this relationship.

We'll now introduce two newer machine learning techniques that are often associated with the field of AI: decision trees and neural networks. These two techniques share a focus on the process of learning, which distinguishes them from regression modeling, where the focus is on fitting mathematical models to data.

Decision Trees

Decision trees are a fairly simple machine learning method and are often used to introduce machine learning models. Their origins date back to the 1930s, but they didn't become popular until the mid-1980s.[9] A decision tree is a set of instructions for guessing the value of some outcome variable by consulting the values of the input variables one by one. A toy example in the domain of criminal justice is shown in figure 1.3. This decision tree provides a way of guessing whether a prisoner up for bail will reoffend (the outcome variable), based on whether they have behaved well in prison and whether they committed a violent offense (two toy input variables).

The tree states that if the prisoner didn't behave well, they will reoffend, regardless of whether their original offense was violent. If they did behave well, they will reoffend if their original offense was violent, and they won't if it wasn't. (We use this crude example for illustration, but in real life they will take many more variables into account.)

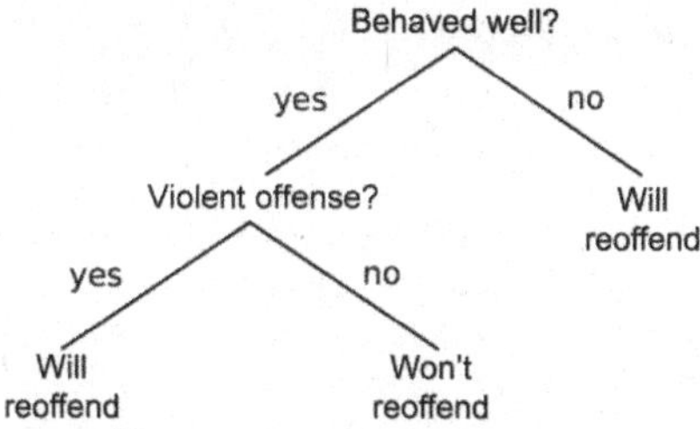

Figure 1.3
A simple decision tree for predicting a prisoner's reoffending.

The key task in decision tree modeling is to devise an algorithm that *creates* a good decision tree from the training data it's given.* In the classic algorithm, we build the decision tree progressively, starting from the top. At each point in the tree, we find the input variable that supplies the most "information" about the outcome variable in the training set and add a node consulting that variable at that point.

An attractive feature of decision trees is that the procedure for reaching a decision can be readily understood by humans. At base, a decision tree is just a complex "if–then" statement. Understandability is an important attribute for machine learning systems making important decisions. However, modern decision tree models often use multiple decision trees embodying a range of different decision procedures and take some aggregate over the decisions reached. "Random forests" are the dominant model of this kind at present. For various reasons, these aggregate methods are more accurate. There is often (although not always) a tradeoff between a machine learning system's explainability and its predictive performance.

Decision trees provide a useful opportunity to introduce the concept of a "classifier," which is widely used in machine learning. A classifier is simply a predictive model whose outcome variable can take a number of discrete values. These discrete values represent different classes that the input items can be grouped into. Decision trees have to operate on variables with discrete values so they can easily implement classifiers. Our decision tree for

* For example, you can imagine the sort of data that would lead an algorithm to create the decision tree in figure 1.3: data on good behavior, episodes of violence, and so on.

reoffending can be understood as a classifier that sorts prisoners into two classes: "will reoffend" and "won't reoffend." To implement classifiers with regression techniques, we must use logistic regression models, which are specifically designed to handle discrete outcome variables.

Neural Networks

Neural networks (sometimes called "connectionist" networks) are machine learning techniques that are loosely inspired by the way brains perform computation. The first neural network was developed by Frank Rosenblatt in 1958,[10] drawing on ideas about neural learning processes developed by Donald Hebb in the late 1940s.[11]

A brain is a collection of neurons, linked together by synapses. Each neuron is a tiny, very simple processor. The brain can learn complex representations and produce complex behavior because of the very large number of neurons it has and the even larger number of synapses that connect them together. Learning in the brain happens through the adjustment of the "strength" of individual synapses—the strength of a synapse determines how efficiently it communicates information between the neurons it connects. We are still far from understanding how this learning process works and how the brain represents information, but neural networks have been very roughly based on this model of the brain, imperfect as it is.

A neural network is a collection of neuron-like units that perform simple computations and can have different degrees of activation. These units are connected by synapse-like links that have adjustable weights. There are many different types of neural networks, but networks that learn a predictive model are predominantly "feedforward networks" of the kind illustrated (very crudely) in figure 1.4. A feedforward network used to learn a predictive model has a set of input units that encode the input variables of training items (or test items), a set of output units that encode the outcome variable for these same items, and a set of intermediate "hidden units." Activity flows from the input units through the hidden units to the output units. Through this process, the network implements a function from input variables to the outcome variable, much like a regression model. Often there are many "layers" of hidden units, each connected to the layer before and the layer after. (The network in figure 1.4 has one hidden layer.)

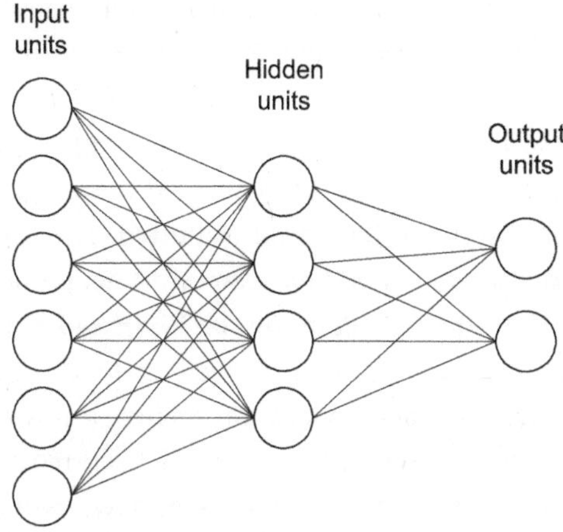

Figure 1.4
A simple feedforward network.

To give a simple example, imagine the network in figure 1.4 is a very simple image classifier that takes a tiny image comprising six pixels, and decides whether these pixels represent an image of type A or type B. The intensity of each pixel would be encoded in the activity of one of the input units. The activity of the output units encodes the type, in some designated scheme. (For instance, type *A* could be encoded by setting the activity of one unit to one, and the other unit to zero, whereas type B could be encoded by setting the activity of the former unit to zero, and the latter unit to one.)

There are many different learning algorithms for feedforward networks. But the basic principle for all of them is "supervised learning." In this learning algorithm, we begin by setting the weights of all the links in the network to random values. The network then implements a function from inputs to outputs in successive rounds of training. To train an image classifier in this fashion, we would first present it with training inputs and let it guess—essentially randomly—what it was "seeing." The classifier will invariably produce errors, and a crucial step in the training process involves calculating what these errors are. This is done by comparing, after each guess, the network's *actual* output values to those it *should* have produced. Errors are then reduced by making small changes to the weights of all links in the network—changes which gradually improve the network's performance as the process is repeated many times.

All the smarts in a supervised learning algorithm relate to how to tweak its weights so as to reduce error. A big breakthrough in this area is the technique of "error backpropagation," which was invented (or at least made prominent) in 1986 by David Rumelhart and colleagues at the University of California, San Diego. This algorithm allowed the weights of neurons in a network's hidden layer(s) to be sensibly adjusted. The invention of backpropagation led to a wave of academic interest in neural networks, but not to immediate practical effect. Then in the late 1990s and early 2000s, various innovations to the feedforward network were devised that culminated in dramatic improvements to backpropagation.[12] These innovations, combined with the huge increases in computing power that occurred around that time, led to a new generation of "deep networks" that have revolutionized AI and data science.

The most salient feature of deep networks is that they have many layers of hidden units (unlike the single layer in figure 1.4). There are now several varieties of deep networks deployed in many different areas of machine learning. Deep networks of one sort or another are often the best performing models. The field of machine learning has in fact undergone a paradigm shift. The majority of researchers in this area currently focus their attention on deep networks. There are several open-source software packages that support the implementation, training, and testing of deep networks (e.g., Google's TensorFlow and Facebook's PyTorch). These packages have undoubtedly helped consolidate the new paradigm, and their ongoing development helps progress it further.

Nevertheless, a significant drawback of deep networks is that the models they learn are so complex that it's essentially impossible for humans to understand their workings. Humans have a reasonable chance of being able to understand a decision tree (or even a set of decision trees) or to understand a regression model that succinctly states the relationships between variables. But they have no chance of understanding how a deep network computes its output from its inputs. If we want our machine learning tools to provide human-understandable explanations of their decisions, we need to supplement them with additional tools that generate explanations. The development of "explanation tools" is a growth area of AI and sufficiently important that we'll discuss explanation systems separately in chapter 2.

Protocols for Testing Predictive Models

So far we've said little about methods to test and evaluate algorithms. Here we'll mention just one device that can be used to evaluate an algorithm's performance.

Often, a predictive algorithm can make several different types of error that have very different implications for its use in the field. Consider a "binary classifier" that's trained to recognize members of one particular class. During testing, this classifier labels each test individual either as "positive" (a member of the class in question), or "negative" (not a member). If we also know the actual class of the test individuals, we can chart how often it's right and wrong in its assignment of these labels and express these results in a "confusion matrix." An example of a confusion matrix is shown in table 1.1. The classifier in this case is a system trained to predict fraudsters. It makes a "positive" response for people it predicts will commit fraud and a "negative" response for everyone else.

The confusion matrix shows how frequently the system is right or wrong in both kinds of prediction. A "false positive" is a case in which the system wrongly predicts someone to commit fraud (a false alarm, so to speak); a "false negative" is a case in which it fails to detect an actual fraudster (a miss). If a system isn't perfect, there will always be a tradeoff between false positives and false negatives (and between true positives and true negatives). For example, an algorithm that judges everyone to be a fraudster will have no false negatives, while one judging nobody to be a fraudster will have no false positives. Importantly, in different domains, we might want to err on one side or the other. For instance, if we're predicting suitability for a rehabilitation project, we might want to err on the side of false positives, whereas if we're predicting guilt in a criminal case, we might want to err on the side of false negatives (on the assumption that it's better to let many more guilty

Table 1.1
Confusion Matrix for a Fraud Detection Algorithm

	Did commit fraud	Did not commit fraud
Predicted to commit fraud	True positives	False positives (type 1 errors)
Predicted not to commit fraud	False negatives (type 2 errors)	True negatives

persons go free than to imprison a single innocent one). For many applications, then, we'd like the evaluation criterion for a classifier to specify *what counts as acceptable performance* in relation to the confusion matrix—that is, what sorts of error we would prefer the system to make.

A confusion matrix is just a simple means of keeping track of what types of errors an algorithm is making and assessing whether or not it's performing acceptably. But there's obviously a lot more to testing and evaluating algorithms than confusion matrices. For one thing, if it's not to be a merely perfunctory, rubber-stamping exercise, testing has to be *frequent*. People's habits, preferences, and lifestyles can be expected to change over time, so the items used to train an algorithm need to be constantly updated to ensure they're truly representative of the people on whom it'll be used. If test items aren't routinely updated, the performance of an algorithm will deteriorate, and biases of various kinds may be (re)introduced. Another requirement is for testing to be *rigorous*. It's not enough to test an algorithm's performance on standard cases—cases involving "average" people on "average" incomes, for example. It must be tested in borderline cases involving unusual circumstances and exceptional individuals.

Mind you, frequent, rigorous testing isn't just important for safety and reliability; it's also crucial to fostering trust in AI. We humans are a suspicious lot, but we're generally willing to go along with a technology if repeated testing under adverse conditions has shown that it's safe to do so. That's why we're happy to board planes. The tests to which Boeing subjects its commercial airliners are famously—even eccentrically—tough-going. The same should go for AI. Until autonomous vehicles clock up enough evidence of safety from rigorous stress-testing in all sorts of conditions (heavy traffic, light traffic, wet weather, with and without pedestrians, with and without cyclists, etc.), it's unlikely we'll be as willing to hand over control to a Google car as we are to an air traffic control system or autopilot.

Summing Up

We've just presented the main varieties of predictive models currently used by government departments and corporations around the world. Our aim has been to emphasize the continuity of modeling techniques. Today's models are extensions of predictive models that have been in use since the dawn of the computer age, and in some cases well before that. We have also emphasized that although these models are often referred to in current discussions as "AI" models, they are often equally well described as "statistical" models. The main novelty of modern AI predictive models is that they often perform better than traditional models, partly because of improvements in techniques, and partly owing to the many new data sources that are coming online in the age of big data. As a result, they're becoming more widely adopted.

In the case of modern AI technologies, it has to be said that commercial companies have taken the lead in development and adoption. The big tech companies (Google, Facebook, Amazon, Facebook, and their Chinese equivalents) are the world leaders in AI, employing the largest, best-qualified and best-funded teams of researchers with access to the largest datasets and greatest computer processing power ever. Academic AI research is lagging some distance behind on all of these metrics. Government departments are lagging even further behind, but they *are* deploying the same kinds of AI tools as the big tech companies. And in both cases, the focus is very much on prediction.

We should emphasize that we haven't covered all of AI, or even all of machine learning, in this chapter. There are many other important machine learning technologies that contribute to modern AI systems, in particular, unsupervised learning methods, which look for patterns in data without guidance, and reinforcement learning methods, where the system's guidance takes the form of rewards and punishments. (Readers interested in these methods can read about them in the appendix to this chapter.) But predictive models as we have defined them are still at the center of the AI technologies that are impacting our world. And as we will show in this book, they are also at the center of citizens' discussions about how AI technologies should be regulated. For instance, the fact that predictive models can all be evaluated in the same basic way makes them a natural focus for discussions about quality control and bias (chapter 3); the fact that they perform clearly defined tasks makes them a natural focus for discussions about human oversight and control (chapter 5); and the fact that their performance is often achieved at the expense of simplicity makes them a natural focus for discussions about transparency (chapter 2). In summary, the class of predictive models is coherent both from a technical perspective and from the perspective of regulation and oversight: it is thus a good starting point for a citizen's guide.

Appendix: Other Machine Learning Systems

Our introduction to machine learning has focused on "predictive models" and the history of their use in industry and government. Most of the discussion in this book will be focused on these tools. However, there are other types of machine learning that find widespread application alongside predictive models—and these should have some introduction too.

One of these is called "unsupervised" machine learning. Although supervised learning algorithms are told how they should map their input variables onto outcome variables, unsupervised algorithms are given no instruction. They simply look for patterns in large datasets that human analysts might miss. Unsupervised algorithms are often used to build profiles of "typical" customers or "typical" citizens. This can be very helpful in simplifying a complex dataset and re-expressing it in terms of the typical patterns it contains. If we know that people who buy dog biscuits also frequently buy flea collars and visit websites aimed at pet owners, we can create a rough and ready class of people who do all these things. This "customer segmentation," among other uses, makes potential customers easier to target with advertising.

It's hard to measure the performance of unsupervised learning systems because they don't have a specific task to do. In the above example, they simply find groupings of customers based on shared characteristics. This makes the regulatory objectives for these systems hard to define. However, unsupervised learning systems are sometimes used in conjunction with predictive models. For instance, the inputs to a predictive model are sometimes simplified classes created by unsupervised learning algorithms. Let's say we want to construct a classifier that sorts people into two groups: high credit risk and low credit risk. But let's also assume we don't know what features distinguish low-risk from high-risk borrowers. Before constructing the classifier, we can run an unsupervised learning algorithm and see if it detects this particular grouping. If it does, it may reveal features that all low-risk borrowers share (e.g., high savings and medium income) and that all high-risk customers share (e.g., low savings and low income). In this case, we can measure the unsupervised system indirectly by seeing whether it improves the performance of the predictive system. Here, regulatory objectives are easier to formulate. The law could mandate that an unsupervised algorithm meet a certain benchmark in respect of predictive accuracy for credit scoring.

A final type of machine learning worth mentioning is "reinforcement learning." This form of learning is used to train an agent-like system that acts in some environment, the state of which it can sense in some ways. The system takes as its input the current state of this environment, as revealed by its senses, and produces as its output an *action*. Typically, it produces a *sequence* of actions, with each action bringing about a change in its environment. Its stream of perceptions is thus influenced by the actions it performs. A good example of this kind of system is an AI agent playing a video game. As input it might take the current pixel values of the game display, and as output it might generate a reaction by a character in the game.

A system trained by reinforcement learning is like a predictive model, in that it learns to map "inputs" onto "outputs." But there are two main differences. First, it doesn't just "predict" an output; it actually *performs* its prediction, bringing about a change in its environment. Second, it doesn't just make one-off, isolated predictions; it makes a stream of predictions, each one related to the new state it finds itself in.

Reinforcement learning is like supervised learning, in that the system learns from some external source how to map inputs onto outputs. But in reinforcement learning, there's less spoon-feeding. The agent isn't told at each step what it should do. Instead, the agent stumbles upon *rewards* and *punishments* in the environment it's exploring. These stimuli are often sparse, as they are in the real world. Thus, it often has to learn *sequences* of actions that achieve rewards, or that avoid punishments, some distance into the future. A lot of learning in humans, and other animals, happens through this kind of reinforcement, rather than direct supervision. Our supervisor is the school of hard knocks.

Under the hood, a system trained by reinforcement learning is still a kind of predictive model, in our terms. And these days, the predictive model is typically a deep network, trained using backpropagation at each time step. But in reinforcement learning, it's the system's job to work out *how to train* this model—that is, it has to come up with its own set of training examples, mapping inputs onto "desired" outputs. This has implications for the way reinforcement learning systems are evaluated. The system developer doesn't typically have a "test set" of unseen input–output mappings on which performance can be rated. Instead, as learning progresses, the developer evaluates the system by simply seeing how good the system gets at maximizing reward and minimizing punishment.

2 Transparency

There's a famous story about a German horse who could do arithmetic.[1] At least, that's how it seemed at first. His name was Hans—"Clever Hans" they called him—and his legend holds important lessons for us today. Hans was reputedly able to add, subtract, divide, multiply, and even read, spell, and tell the time.[2] If his owner asked him what day of the month the coming Friday would fall on, Hans would tap out the answer—eleven taps for the eleventh day of the month, five taps for the fifth day of the month, just like that. The astonishing thing is that his answers were very often right. He had something like an 89 percent hit rate.[3] Hans understandably caused a stir. He even featured in the *New York Times* under the headline, "Berlin's Wonderful Horse: He Can Do Almost Everything but Talk."[4]

Alas, the adulation wasn't to last. A psychologist investigating his case concluded that Hans wasn't actually performing arithmetic, telling the time or reading German. Rather less sensationally—but still quite remarkably—Hans had become adept at reading his *owner*. It was revealed that as Hans would approach the right number of taps in response to any of the questions put to him, his owner gave unconscious postural and facial cues that Hans was able to pick up on. The owner's tension would be greatest at the second-to-last tap before the correct tap, after which the owner's tension eased. Hans would stop tapping exactly one tap after his owner's tension peaked. So Hans didn't know the answers to the questions he was asked. He gave the right answers alright, but for the wrong reasons.

The lesson resonates with anyone that's guessed their way through a multiple-choice quiz and ended up with a high score, but it's also illustrated vividly today by AI. The risk of a machine learning tool yielding correct results for frankly spurious reasons is high—indeed, worryingly high. In the prologue, we already mentioned how an object classifier trained on pictures of wolves with snow in the background is likely to discriminate between a wolf and a husky based on that single fact alone rather than on features of the animal it's supposed to be recognizing, like the eyes and snout. But this isn't an isolated case. Some classifiers will detect a boat only if there's water around,[5] a train only if there's a railway nearby,[6] and a dumb-bell only if there's an arm lifting it.[7] The moral of the story is simple: we should never trust a technology to make important decisions about anything unless we've got some way of interrogating it. We have to be satisfied that its "reasons" for deciding one way or another actually check out. This makes transparency something of a holy grail in AI circles. Our autonomous systems must have a way for us to be able to scrutinize their operations so we can catch any Clever Hans-type tricks in good time. Otherwise a classifier that works well enough in standard contexts (when there's lots of water around, say, or lots of other gym equipment strewn across the floor) will fail abysmally in nonstandard contexts. And let's face it, the world of high-stakes decision-making nearly always takes place in nonstandard contexts. It's not so much whether a system can handle the standard open-and-shut case that we care about when the Home Office assesses a visa application—it's precisely the *outlier* case that we're most anxious about.

But now what exactly does it mean for a system to be transparent? Although almost everyone agrees that transparency is nonnegotiable, not everyone's clear about what exactly we're aiming for here. What sort of detail do we require? Is there such a thing as too much detail? At what point can we be satisfied that we understand how a system functions, and why it decided *this* way rather than *that*?

Questions like these are vitally important in the brave new world of big data, so we'd like to consider them carefully in this chapter. First, though, let's lay out the terrain a little. After all, transparency covers a *lot* of ground—and here we're only interested in a small patch of it.

The Many Meanings of "Transparency"

"Transparency" can mean a lot of different things. It has general meanings, as well as more specific meanings. It can be aspirational and whimsical but also definite and concrete.

At the broadest level it refers to *accountability* or *answerability*, that is, an agency's or person's responsiveness to requests for information or willingness to offer justification for actions taken or contemplated. This is the most overtly political sense of the term. Importantly, it's a *dynamic* sense—there's never a moment when, as an authority committed to this ideal, your commitment is fully discharged. To be committed to transparency means *being* transparent, not *having been* transparent. We expect our elected representatives to act in the public interest, and transparency stands for their ongoing obligation to meet this expectation. When government is open, answerable, and accountable to its citizens, it is less tempted to become insular, self-serving, and corrupt. Transparency in this broad sense is therefore a safeguard against the abuse of power. All democracies notionally value this sense of transparency, but it's obviously hazy and aspirational. From here, the notion branches out in at least three directions, each of which takes the concept onto much firmer ground.

In one direction, transparency may be associated with moral and legal responsibility (see chapter 4). This captures such familiar notions as blameworthiness and liability for harm. Here the sense of transparency is often static, i.e., "once-for-all" or "point-in-time" (e.g., "judgment for the plaintiff in the amount of $600). Unlike the broader notion we started with, this sense of transparency isn't necessarily dynamic. Rather than prospectively *preventing* wrongdoing, it's most often *corrective* (and retrospective). But, as we point out in chapter 4, responsibility isn't always backward looking and static in this way. Companies that undertake to manufacture goods in a form that will reach the end-user *as is*—that is, without further possibility of the goods being road-tested or otherwise treated—will have a forward-looking responsibility to make sure the goods are safe. This is what lawyers mean by a "duty of care," and it is one example where legal responsibility has some of the qualities we tend to associate with accountability and answerability.

In a second direction, transparency definitely retains its dynamic quality but relates more narrowly to the inspectability (or auditability) of institutions, practices and instruments. Here transparency is about mechanisms: How does this or that tool actually *work*? How do its component parts fit together to produce outcomes like those it is designed to produce? Algorithms can be "inspected" in two ways. First, we can enquire of their provenance. How were they developed, by whom, and for what purpose(s)? This extends to procurement practices. How were they acquired, who commissioned them, on what terms, and—the lawyer's question—*cui bono*, that is, who benefits? This might be called *process* transparency. Second, we can ask of any algorithm, how does it work, what data has it been trained on, and by what logic does it proceed? This might be called *technical* transparency, and centers on the notion of *explainability*. Before any particular decision is reached using an algorithm, we can seek *general* ("*ex ante*") explanations. For instance, in a machine learning case, we can ask whether we're dealing with a decision tree, a regression algorithm, or some mixture. Information about the kind of algorithm we're dealing with can tell us quite a lot about its general principles of operation, and whether algorithm A is better than algorithm B. In the wake of any *particular* algorithmic decision, however, the questions posed can be more specific. Why did the algorithm decide *this* matter in *this* particular way? What are its "reasons" for so deciding? This is to seek a specific, individualized ("*ex post*" or "*post hoc*") explanation. In both of these cases it's important to remember that just because a decision system is explainable doesn't mean that all interested parties will be in a position to understand the explanation. If you want to challenge an automated decision at law, at the very least your lawyers will need to understand something about how the underlying algorithm works. This makes *intelligibility* a further key feature that any explainable system ought to have, and by this, of course, we mean intelligibility *relative to some domain of expertise*. Obviously, intelligibility for legal purposes would be different from intelligibility for software engineering purposes; coders and lawyers will want to know and be able to understand different things about an automated system. A final property that it's desirable for an explainable system to have—especially when we're talking about explanations of automated decisions—is *justifiability*. We don't just want explanations, or even intelligible ones. We also want good, fair, and reasonable explanations based on plausible reasoning.

In a third direction, transparency denotes accessibility. Meaningful explanations of an algorithm may be possible, but they may not be *available*. Intellectual property rights might prevent the disclosure of proprietary code or preclude access to training data so that, even if it were possible to understand how an algorithm operates, a full reckoning may not be possible for economic, legal, or political reasons. Algorithms that are otherwise technically transparent may therefore be "opaque" for nontechnical reasons. Figure 2.1 depicts these various nested and interacting notions of transparency diagrammatically.

In the context of algorithms and machine learning, concerns have been raised about transparency in every one of these senses. The sense which has

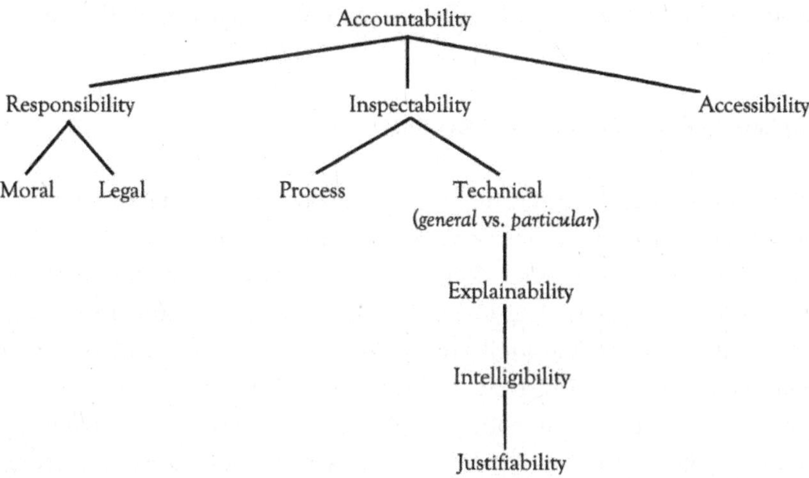

Figure 2.1
The various senses of transparency.

much exercised civil libertarians and a growing number of computer scientists, however, is technical transparency—that is, *explainability* and *intelligibility*. This is what we'll focus on in this chapter. It's the explainability of a system's decisions that helps us sort out the clever from the merely Clever Hans. (Process transparency is also a major concern, but we'll cover that in coming chapters.)

First up, we'll consider why algorithms have a "technical" transparency problem. Why are some algorithms' decisions so difficult to explain and interpret that we must consider them opaque? Second, we'll consider just how much of a problem this really is. We'll set automated decision-making against human decision-making to see how much of the problem is unique to automated systems and how much of it afflicts *all* decision makers, humans included. This is important because if human decision-making poses a similar transparency problem to the one posed by automated decision-making, perhaps we shouldn't be too condescending when discussing the opacity of machine learning systems, insisting on lofty standards of transparency for machines that not even we humans would be capable of meeting.

As you can probably tell, we think it's reasonable for the standards we apply to ourselves when assessing human decisions to set some sort of benchmark when assessing automated decisions. Artificial intelligence, after all, aspires to human-level intelligence in a variety of domains, and these not infrequently involve some form of routine decision-making.

Explanations and Reasons for Decisions

When someone is charged with pronouncing on your personal affairs—your entitlements, your rights, your obligations—we take it for granted that they should be able to justify their decision, even if they don't, strictly speaking, *have* to. We naturally assume, for example, that if a bank declines to give you a loan, the bank didn't just roll a die or flip a coin. If a tenancy tribunal says your rent increase was justified, we expect there to be reasons—in this case laws—behind this determination. Explanations for decisions, whether by courts, public officials, or commercial entities, may not always be forthcoming. Private businesses generally have no obligation to explain themselves to you, and even public agencies aren't always required to furnish reasons for their decisions. But we—all of us—assume that explanations *could* be given, in principle. This is because decision-making shouldn't be arbitrary or capricious. As we noted earlier, it should be based on plausible reasoning.

Of course, explanations sometimes *are* mandatory in public decision-making. This is most clear in legal contexts, where courts of law are generally obliged to provide reasons for their decisions. The obvious rationale for this requirement is to facilitate appeal. Provided there *is* a right to appeal, you need reasons. How are you supposed to appeal a judgment unless you know how the court came to its decision? Generally, the moment a true right of appeal exists, a decision maker's duty to give reasons is implied. On the other hand, if you don't have a right of appeal, there's no strict need for a court to provide reasons. (See box 2.1.)

Of course, the duty to provide reasons may still be imposed by law for other reasons, for example, to further the aims of open justice. Often knowing the reasons why a particular decision has been taken, even if only in rough outline, can engender trust in the process that led to it and confidence that the people in charge of the process acted fairly and reasonably. Thus explanations can have both instrumental value as a means to overturn an adverse determination by appeal and intrinsic value as a democratic index of accountability and transparency in the broadest sense we considered above. In many legal systems, too, statements of reasons function as a source of law—we call it "precedent"—through which, over time,

Box 2.1

Administrative Law and the Difference between "Appeal" and "Review"

Administrative law is the area of law concerned with the actions of public officials (ministers, secretaries, attorneys-general, directors-general, commissioners, etc.). Traditionally, in common-law countries (UK, Canada, Australia, and New Zealand), there was no general duty for public officials to provide reasons for their decisions precisely because there was no true right of appeal from their decisions. Instead, there was a right to *review* (sometimes called "judicial review"), which can be understood as something like the right to appeal the *circumstances* of a decision, but not the decision itself. For example, if a public official is given authority (by legislation) to determine land rates within a municipality but not land *taxes*, one might challenge the official's decision to alter land taxes. Here you aren't said to be *appealing* the decision so much as having it *reviewed*. You don't need a published statement of reasons to review a decision in this sense because you aren't challenging the decision itself (i.e., how the official reasoned from the facts and the law to their conclusion). You're only challenging the *circumstances* of the decision (i.e., the fact that the official did not have power to fix land taxes in the first place). True appeals, in contrast, bring the decision itself into question, not just the circumstances of the decision. This sometimes gets called "merits review." It's the right that litigants frequently exercise in criminal and civil proceedings. A true right of appeal allows the appeal court to examine, in detail, the reasons for the decision maker's determination, including how they assessed the evidence, what conclusions they drew from it, how they interpreted the law, and how they applied that interpretation to the facts of the case. In other words, an appeal allows the appeal court to substitute its own decision for that of the original decision maker. For a true right of appeal to be exercised, clearly the aggrieved party needs a statement of reasons!

consistency in the law is achieved by treating like cases alike. And obviously the habit of providing reasons promotes better quality decision-making all round. A judge conscious of the many hundreds (or even thousands) of lawyers, students, journalists, and citizens poring over their judgment will be anxious to get it right—their reputation is on the line.

Someone's "right" to an explanation is, of course, someone else's "duty" to provide it. Understandably, jurisdictions differ on the existence of this right. Some jurisdictions (like New Zealand) have imposed this duty on public officials.[8] Others (like Australia) have declined to do so (except in the case of judges). Interestingly, the EU is potentially now the clearest case of a jurisdiction requiring explanations for automated decisions in both the public *and* private sectors (see below). Still, it's debatable whether explanation rights provide an adequate remedy in practice. Although they certainly have a place, their effectiveness can easily be overstated, as they do rely on individuals being aware of their rights, knowing how to enforce them, and (usually) having enough money to do so.[9]

In any case, it's hardly surprising that, as algorithmic decision-making technology has proliferated, civil rights groups have become increasingly concerned with the scope for challenging algorithmic decisions. This isn't just because AI is now regularly being recruited in the legal system. AI is also being used by banks to determine creditworthiness, by employment agencies to weed out job candidates, and by security companies to verify identities. Although we can't "appeal" the decisions of private businesses the way we can appeal court decisions, still there are laws that regulate how private businesses are to conduct themselves when making decisions affecting members of the public. Anti-discrimination provisions, for example, prevent the use of a person's ethnicity, sexuality, or religious beliefs from influencing the decision to hire them. A company that uses AI probably purchased the software from another company. How can we be sure the AI doesn't take these prohibited ("protected") characteristics into account when making a recommendation to hire a person or refuse a loan? Do we just *assume* the software developers were aware of the law and programmed their system to comply?

AI's Transparency Problem

Traditional algorithms didn't have a transparency problem—at least not the same one that current deep learning networks pose. This is because traditional algorithms were defined "by hand," and there was nothing the system could do that wasn't already factored into the developer's design for how the system should operate given certain inputs.*

As we saw in chapter 1, however, deep learning is in a league of its own. The neural networks that implement deep learning algorithms mimic

*Of course, some traditional algorithms, like "expert systems," *could* also be inscrutable in virtue of their complexity.

the brain's own style of computation and learning. They take the form of large arrays of simple neuron-like units, densely interconnected by a very large number of plastic synapse-like links. During training, a deep learning system's synaptic weights are adjusted so as to improve its performance. But although this general *learning algorithm* is understood (e.g., the "error backpropagation" method we discussed in chapter 1), the actual *algorithm learned*—the unique mapping between inputs and outputs—is impenetrable. If trained on a decision task, a neural network basically derives its own method of decision-making. And there is the rub—it is simply not known in advance what methods will be used to handle unforeseen information. Importantly, neither the user of the system nor its developer will be any the wiser in this respect. *Ex ante* predictions and *ex post* assessments of the system's operations alike will be difficult to formulate precisely. This is the crux of the complaint about the lack of transparency in today's algorithms. If we can't ascertain exactly why a machine decides the way it does, upon what bases can its decisions be reviewed? Judges, administrators, and public agencies can all supply reasons for their determinations. What sorts of "reasons" can we expect from automated decision systems, and will such explanations be good enough? What we're going to suggest is that if human decision-making represents some sort of gold standard for transparency—on the footing that humans readily and routinely give reasons for their decisions—we think AI can in some respects already be said to meet it.

What Kinds of Explanations Have Been Demanded of Algorithmic Systems?

To date, calls for transparent AI have had a particular ring to them. Often there's talk of inspecting the "innards" or "internals" of an algorithmic decision tool,[10] a process also referred to as "decomposition" of the algorithm, which involves opening the black box to "understand how the *structures within*, such as the weights, neurons, decision trees, and architecture, can be used to shed light on the patterns that they encode. This requires access to the bulk of the model structure itself."[11] The IEEE raises the possibility of designing explainable AI systems "that can provide justifying reasons or other reliable 'explanatory' data *illuminating the cognitive processes* leading to ... their conclusions."[12] Elsewhere it speaks of "internal" processes needing to be "traceable."[13]

It isn't just aspirational material that's been couched in this way. Aspects of the EU's General Data Protection Regulation (GDPR) have engendered similar talk.[14] For instance, the Article 29 Data Protection Working Party's draft guidance on the GDPR states that "a complex mathematical explanation about how algorithms or machine-learning work," though not generally relevant, "should also be provided if this is necessary to allow experts to further verify how the decision-making process works."* There's a sense in which this verges on truism. Obviously technical compliance teams, software developers and businesses deploying algorithmic systems may have their own motivations for wanting to know a little more about what's going on "under the hood" of their systems. These motivations may be perfectly legitimate. Still, it's a fair bet that such investigations won't be concerned with AI decisions *as* decisions: they'll be concerned with the technology as a piece of kit—an artifact to be assembled, disassembled and reassembled, perhaps with a view to it making better decisions in due course, bug-fixing, or enhancing human control. When, on the other hand, we want to know why a system has *decided* this way or that, and hence seek *justifying* explanations, we think that most often—although not in every case—the best explanations will avoid getting caught up in the messy, technical internal details of the system. The best explanations, in other words, will resemble human explanations of action.

Human Explanatory Standards

And just what sorts of explanations are human agents expected to provide? When judges and officials supply written reasons, for example, are these expected to yield the entrails of a decision? Do they "illuminate the cognitive processes leading to a conclusion"? Hardly.

It's true that human agents are able to furnish reasons for their decisions, but this isn't the same as illuminating cognitive processes. The cognitive processes underlying human choices, especially in areas in which a crucial element of intuition, personal impression, and unarticulated hunches are

*Strictly speaking, this "good practice" recommendation (Annex. 1) relates to Article 15, not Article 22, of the GDPR. Article 15(1)(h) requires the disclosure of "meaningful information about the logic involved" in certain kinds of fully automated decisions.

driving much of the deliberation, are in fact far from transparent. Arenas of decision-making requiring, for example, assessment of the likelihood of recidivism or the ability to repay a loan, more often than not involve significant reliance on what philosophers call "subdoxastic" factors—factors beneath the level of conscious belief. As one researcher explains, "a large part of human decision making is based on the first few seconds and how much [the decision makers] like the applicant. A well-dressed, well-groomed young individual has more chance than an unshaven, disheveled bloke of obtaining a loan from a human credit checker."[15] A large part of human-level opacity stems from the fact that human agents are also frequently *mistaken* about their real (internal) motivations and processing logic, a fact that is often obscured by the ability of human decision makers to invent *post hoc* rationalizations. Often, scholars of explainable AI treat human decision-making as privileged.[16] Earlier we noted that some learning systems may be so complex that their manipulations defy systematic comprehension and that this is most apparent in the case of deep learning systems. But the human brain, too, is largely a black box. As one scholar observes:

> We can observe its inputs (light, sound, etc.), its outputs (behavior), and some of its transfer characteristics (swinging a bat at someone's eyes often results in ducking or blocking behavior), but we don't know very much about how the brain works. We've begun to develop an algorithmic understanding of some of its functions (especially vision), but only barely.[17]

No one doubts that well-constructed, comprehensive, and thoughtful human reasons are extremely useful and generally sufficient for most decision-making purposes. But in this context, usefulness and truth aren't the same. Human reasons are pitched at the level of what philosophers call "practical reason"—the domain of reason that concerns the justification of action (as distinguished from "epistemic" or "theoretical reason," which concerns the justification of belief). Excessively detailed, lengthy, and technical reasons are usually not warranted, or even helpful, for most practical reasoning. This doesn't mean that the structure of typical human reasoning is ideal in all circumstances. It means only that for most purposes it will serve adequately.

Consider decisions made in the course of ordinary life. These are frequently made on the approach of significant milestones, such as attaining the age of majority, entering into a relationship, or starting a family, but they most often involve humdrum matters (should I eat in, or go out for dinner tonight?). Many of these decisions will be of the utmost importance to the person making them and may involve a protracted period of deliberation (e.g., what career to pursue, whether to marry, whether and when to have children, and what to pay for a costly asset—a home, a college education, etc.). But the rationales that may be expressed for them later on, perhaps after months of research or soul-searching, will not likely assume the form of more than a few sentences. Probably there will be one factor among three or four that reveals itself after careful reflection to be the most decisive, and the stated *ex post* reasons for the decision will amount to a statement identifying that particular factor together with a few lines in defense.

Actually, if you think about it, most "official" decision-making is like this too. It might concern whether to purchase new equipment, whether to authorize fluoridation of a town's water supply, whether to reinstate someone unfairly dismissed from a place of employment, whether to grant bail or parole, or whatever. But basically, the decision's formal structure is the same as that of any other decision, public, personal, commercial, or otherwise. True, the stakes may be higher or lower, depending on what the decision relates to and how many people will be affected by it. Also, the requirement to furnish reasons—as well as the duty to consider certain factors—may be mandated in the one case and not the other. But the primary difference isn't at the level of form. Both contexts involve practical reasoning of a more or less systematic character. And furnishing explanations that are more detailed, lengthy, or technical than necessary is likely to be detrimental to the aims of transparency, regardless of the public or private nature of the situation.

There are, of course, some real differences between public and private decision-making. For example, certain types of reasons are acceptable in personal but not public decision-making. It may be fine to say, "I'm not moving to Auckland because I don't like Auckland," but the same sort of reasoning would be prohibited in a public context. Furthermore, public decision-making often takes place in groups to mitigate the "noisiness" of individual reasoners, such as committees, juries, and appeal courts—although even here, many private, purely personal decisions (regarding, e.g., what to study, which career to pursue, whether to rent or purchase, etc.), are also frequently made in consultation with friends, family, mentors, career advisers, and so on. In any case, these differences don't detract from their fundamentally identical structure. Either way, whether there are more or fewer people involved in the decision-making process (such as jurors, focus groups, etc.) or whether there are rights of appeal, both decision procedures employ practical reasoning and take beliefs and desires as their inputs. Take judicial decision-making—perhaps the most constrained and regimented form of official reasoning that exists. Judicial reasoning is, in the first instance, supposed to appeal to ordinary litigants seeking the vindication of their rights or, in the event of a loss, an explanation for why such vindication won't be forthcoming. So it simply must adopt the template of practical reason, as it must address citizens in one capacity or another (e.g., as family members, corporate executives, shareholders, consumers, criminals, etc.). Even in addressing itself to lawyers, for example, when articulating legal rules and the moral principles underpinning them, it cannot escape or transcend the bounds of practical (and moral) reasoning.[18]

We're not claiming that these insights are in any sense original, but we do think they're important. As we've intimated, the decision tools co-opted in predictive analytics have been pressed into the service of practical reasoning. The aim of the GDPR, for instance, is to protect "natural persons" with regard to the processing of "personal" data (Article 1). Articles 15 and 22 concern a data subject's "right" not to be subject to a "decision" based solely on automated processing, including "profiling." The tools that have attained notoriety for their problematic biases, such as PredPol (for hot-spot policing) and COMPAS (predicting the likelihood of recidivism), likewise involve software intended to substitute or supplement practical human decision-making (for instance, by answering questions such as, how should we distribute police officers over a locality having X geographical characteristics? What is the likelihood that this prisoner will recidivate? Etc.). Explanations sought from such technologies should aim for levels that are appropriate to practical reasoning. Explanations that would be too detailed, lengthy, or technical to satisfy the requirements of practical reasoning shouldn't be seen as ideal in most circumstances.

It is a little odd, then, that many proposals for explainable AI assume (either explicitly or implicitly) that the innards of an information processing system constitute an acceptable and even ideal level at which to realize the aims of transparency. A 2018 report by the UK House of Lords Select Committee on Artificial Intelligence is a case in point. On the one hand, what the Committee refers to as "full technical transparency" is conceded to be "difficult, and possibly even impossible, for certain kinds of AI systems in use today, and would in any case not be appropriate or helpful in many cases."[19] On the other hand, something like full technical transparency is "imperative" in certain safety-critical domains, such as in the legal, medical, and financial sectors of the economy. Here, regulators "must have the power to mandate the use of more transparent forms of AI, even at the potential expense of power and accuracy."[20] The reasoning is presumably that whatever may be lost in terms of accuracy will be offset by the use of simpler systems whose innards can at least be properly inspected. So you see what's going on here. Transparency of an exceptionally high standard is being trumpeted for domains where human decision makers themselves are incapable of providing it. The effect is to perpetuate a double standard in which machine tools must be transparent to a degree that is in some cases unattainable in order to be considered transparent at all, while human decision-making can get by with reasons satisfying the comparatively undemanding standards of practical reason. On this approach, if simpler and more readily transparent systems are available, these should be preferred even if they produce decisions of inferior quality. And so the double standard threatens to prevent deep learning and other potentially novel AI techniques from being implemented in just those domains that could be revolutionized by them. As the committee notes:

> We believe it is not acceptable to deploy any artificial intelligence system which could have a substantial impact on an individual's life, unless it can generate *a full and satisfactory explanation for the decisions it will take*. In cases such as deep neural networks, where it is not yet possible to generate *thorough* explanations for the decisions that are made, this may mean delaying their deployment for particular uses until alternative solutions are found.[21]

This might be a sensible approach in some cases, but it's a dangerous starting position. As the committee itself noted, restricting our use of AI only to what we can fully understand limits what can be done with it.[22] There are various high-stakes domains, particularly in clinical medicine and psychopharmacology, where insisting on a thorough understanding of a technology's efficacy before adopting it could prove hazardous to human wellbeing.

Unconscious Biases and Opacity in Human Decision Making

It's a widely accepted fact that "humans are cognitively predisposed to harbor prejudice and stereotypes."[23] Not only that, but "contemporary forms of prejudice are often difficult to detect and may even be unknown to the prejudice holders."[24]

Recent research corroborates these observations. It seems that the tendency to be unaware of one's own biases is even present in those with regular experience of having to handle incriminating material in a sensitive and professional manner. In a recent review of psycho-legal literature comparing judicial and juror susceptibility to prejudicial publicity, the authors note that although "an overwhelming majority of judges and jurors do their utmost to bring an impartial mind to the matters before them ... even the best of efforts may nonetheless be compromised."[25] They write that "even accepting the possibility that judges do reason differently to jurors, the psycho-legal research suggests that this does not have a significant effect on the fact-finding role of a judge,"[26] and that "in relation to prejudicial publicity, judges, and jurors are similarly affected."[27]

Findings like these should force us to reassess our attitudes to human reasoning and question the capabilities of even the most trusted reasoners. The practice of giving reasons for decisions may be simply insufficient to counter the influence of a host of factors, and the reasons offered for human decisions can well conceal motivations hardly known to the decision makers themselves. Even when the motivations *are* known, the stated reasons for a decision can serve to cloak the true reasons. In common law systems it's well known that if a judge has decided upon a fair outcome, and there's no precedent to support it, the judge might just grope around until *some* justification can be extracted from what limited precedents do exist.[28]

Sometimes in discussions about algorithmic transparency you hear people cite the possibility of *appealing* human decisions, as though this makes a real difference to the kind of transparency available from human decision makers. Although it's understandable to view courts and judges as paradigms of human decision-making, what's often forgotten is that legal rights of appeal are quite limited. They can rarely be exercised automatically. Often the rules of civil procedure will restrict the flow of appeals from lower courts by requiring appeal courts to "grant leave" first.[29]

But the situation is worse than this. Substantial parts of judicial reasoning are effectively immune from appeal (albeit for pragmatic reasons), even in the lowest courts, *and even when appeals are theoretically possible!* A degree of judicial discretion probably has to be exercised in every case, and yet discretion can often only be appealed within severely narrow limits.[30] Given how frequently judges are called on to exercise their discretion, this could be seen as contrary to the principles of open justice. Judges are also allowed considerable leeway in respect of their findings on witness credibility. Appeal courts are generally reluctant to overturn judicial determinations of credibility, because the position of trial judges in being able to assess the demeanor of a witness at first hand is seen to deserve particular respect. And let's not forget that jury deliberations are quintessential black boxes. No one (apart from the jurors themselves) can know why a jury decided the way it did, even when appeals from their verdicts are possible. How's that for transparency!

But let's dig a little deeper into the cognitive underground. The purely neurophysiological aspects of human decision-making are not understood beyond general principles of interneural transmission, excitation, and inhibition. In multi-criterion decision cases in which a decision maker must juggle a number of factors and weigh the relevance of each in arriving at a final decision, one hypothesis suggests that the brain eliminates potential solutions such that a dominant one ends up inhibiting the others in a sort of "winner takes all" scenario.[31] Although this process is to some extent measurable, "it is essentially hidden in the stage where weights or relative importance are allocated to each criterion."[32] It serves as a salutary reminder that even when a sentencing judge provides reasons allocating weights to various statutory factors, the actual inner processing logic behind the allocation remains obscure.

More general work on the cognitive psychology of human decision-making is no less sobering. "Anchoring" and "framing" effects are well known to researchers in the field. One such effect, the "proximity" effect, results in more recent events having greater weight than older ones and bearing a greater influence on choices in the search for solutions.[33] The tendency to see false correlations where none exists is also well documented.[34] This bias is at its strongest when a human subject deals in small probabilities.[35] Finally, constraints imposed by short-term memory capacity mean we can't handle more than three or four relationships at a time.[36] Because it's in the nature of complex decisions to present multiple relationships among many issues, our inability to concurrently assess these factors constitutes a significant limitation on our capacity to process complexity.

The upshot is simple: let's not pretend we humans are paragons of transparency next to those unfathomable black-boxes we call deep networks.

Explainable AI 2.0

We've suggested that because the demands of practical reason require the justification of action to be pitched at the level of practical reason, decision tools that support or supplant practical reasoning generally shouldn't be expected to aim for a standard any higher than this. In practice, this means that the sorts of explanations for algorithmic decisions that are analogous to ordinary, run-of-the-mill, interpersonal explanations should be preferred over ones that aim at the architectural innards of a decision tool. The time has come to flesh this out. What exactly would these analogues of everyday explanations look like?

Modern predictive models operating in real-world domains tend to be complex things, regardless of which machine learning methods they use (see chapter 1). If we want to build a predictive system that can convey to a human user *why* a certain decision was reached, we have to add functionality that goes beyond what was needed to generate the decision in the first place. The development of "explanation tools" that add this functionality is a rapidly growing new area of AI.

The basic insight behind the new generation of explanation tools is that, to understand how one predictive model works, *we can train another predictive model to reproduce its performance.* Although the original model can be very complex and optimized to achieve the best predictive performance, the second model—the "model-of-a-model"—can be much simpler and optimized to offer maximally useful explanations.

Perhaps the most useful thing a decision subject wants to know is how different factors were weighed in coming to a final decision. It's common for human decision makers to disclose these allocations, even if, as we mentioned earlier, the inner processing logic leading to them remains obscure. Weights are classic exemplars of everyday logic, and one way for algorithmic decision tools to be held accountable in a manner consistent with human decision-making is by having them divulge their weights.[37] Some of the most promising systems in this area are ones that build a local model of the factors most relevant to any given decision of the system being explained.[38] Such "model-of-a-model" explanation systems also have the added benefit of being able to explain how a system arrived at a given decision without revealing any of its internal workings. This should please tech firms. By providing explanations of how their software "works," tech companies needn't worry that they'll necessarily be disclosing their patented "secret sauce" at the same time. This feature of model-of-a model systems shouldn't be surprising. Remember, they aren't actually telling you "*this* is how the algorithm decided X, *this* is how the algorithm works." Instead, as their very name implies, they're providing a *model*, and a model only has to give a high-level, simplified description of something. Rather like the London Tube map—its depiction of the London Underground is economical and compact, which is no doubt what makes it useful for catching the tube, but obviously no one thinks it provides a reliable guide to London's topography. It'd be fairly useless for navigating at street level, for example.

Still, you might wonder, can explanation systems go one better and give you explanations that reflect more faithfully how an algorithm actually decides while still being intelligible—indeed, while mimicking the structure of our own humanoid logic? The answer, it seems, is yes. Researchers at Duke University and MIT built an image classifier that not only provides comprehensible human-style explanations, but that actually proceeds in accordance with that very logic when classifying images.[39] Imagine for a moment you're presented with a picture of a bird, and your job is to recognize what species it belongs to—a pretty standard object classification task, you might say. It's not always easy, even for trained ornithologists, to correctly classify a bird just by looking at an image. There are simply too many species to contend with. But let's say you were to have an educated guess. How would you explain your answer? We considered this sort of problem when we discussed the huskies and wolves. Remember we noted there that if you were going to justify why you thought an image of a husky was actually a wolf, there'd be certain things you'd point out: the eyes, the ears, and the snout, most likely. In other words you'd *dissect* the image and point out that this or that part of the image is typical for this and that species of canine. The combined weight of all these "typical" features would point us in the direction of one species over another. In the prologue, we noted that many object classifiers would not reason this way, instead focusing on the absence or presence of snow in the image (which is understandable given the likely number of images of wolves in the training set with snow in the background but still absolutely ridiculous). Well, the researchers at Duke and MIT managed to build a bird species classifier that reasons more or less as we would—by dissecting the bird image and comparing selected parts against species-typical parts in a training set. Importantly, this isn't just how the classifier reasons to a conclusion—it's also how it *explains* its conclusions. As the team put it, "it has a transparent reasoning process that is *actually* used to make predictions."[40] Win-win.

For a few years now, companies have resisted providing explanation systems, taking refuge behind the excuse that it's either too difficult, that the explanations would be incomprehensible, or that disclosure risks compromising trade secrets. Slowly, the winds are changing, and even big players are beginning to see that explainability isn't just important—it's potentially commercializable. There's now "a growing industry of consultancy firms that claim to provide algorithmic auditing 'as a service' in order to assess existing algorithms for accuracy, bias, consistency, transparency, fairness, and timeliness."[41] Google and IBM are also getting in on the act. IBM has launched its very own explanation tool—its cloud-based open-source software will enable customers "to see, via a visual dashboard, how their algorithms are making decisions and which factors are being used in making the final recommendations. It will also track the model's record for accuracy, performance, and fairness over time."[42] And none other than Google itself has launched a "what-if" tool, "also designed to help users look at how their machine-learning models are working."[43]

Double Standards: Good or Bad?

A crucial premise of this chapter has been that the standards of transparency should be applied consistently, regardless of whether we're dealing with humans or machines. Partly this is because the field of AI to some extent makes human achievement a standard worth striving for and partly this is because (we have suggested) humans and machines are both opaque in their own ways. Differences will naturally arise when one system is organic and the other is synthetic, but these differences don't seem to justify adopting different standards of transparency. This isn't to say there are *no* circumstances in which different standards might be required—there certainly are. But they are exceptional (and probably a little too esoteric to get into here).[44] Instead of considering such cases, let's consider a few other factors which probably *don't* justify imposing different standards on humans and machines, although they might seem to do so at first.

One factor we can think of is the potential of AI to advance well beyond the level of human transparency. If algorithmic decision tools have a good chance of being significantly better than humans at explaining themselves, then regulations probably *should* be crafted with a view to bringing out the best that they can be—even if this means setting a regulatory standard that would be far stricter than one we'd ever apply to ourselves. However, we have our doubts about just how much less of a black box a multimillion-neuron deep network is likely to be than a human brain, in *practical* terms anyway. If an artificial intelligence is only *in principle* less opaque than a human intelligence, but not in practice, the two intelligences would be comparably opaque, and a double standard hard to justify.

Another argument for double standards might run as follows. The kind of decisions we're worried about when discussing algorithmic decision-making are decisions regarding policies that affect third parties. In these situations, procedures are in place to minimize individuals' biases, such as expert reports, committees, and appeal mechanisms. And such procedures might be thought to tip the scales in favor of human decision-making, justifying a more lenient standard of transparency.

Now we've already pointed out why appeal mechanisms are restrictive and limited in their potential to reduce bias. Regarding committees, we cited a recent paper demonstrating that both juries (a type of committee) and judges are vulnerable to prejudicial media publicity. So just having more people involved in a decision doesn't necessarily eliminate or reduce the potential for human bias to interfere with human reasoning. As for expertise, judges are a type of expert, and, as we said, even when their own motivations are known, the stated reasons for their decisions can serve to cloak the true reasons. But maybe there's more to this point about committees. Here the thought is that a high standard of transparency is naturally enforced by processes within a group, because members often need to justify and rationalize their points of view, which are typically challenged or queried in the ordinary course of discussion.

But actually, research in social psychology suggests that group-based mechanisms that ensure the *production* of justifications don't always guarantee their *quality*. In fact, participants in a group are often swayed by the mere presence of a justification, regardless of its quality. A classic study found that intrusions into a photocopier queue were more likely to be tolerated if a justification was provided, even if it was devoid of content.[45] "May I use the Xerox machine because I have to make copies?" was more effective than "May I use the Xerox machine?" Of course, the result speaks directly to the dynamics of an informal group setting, not a high-level public committee, but it has been taken seriously by legal theorists in discussions of legitimacy.[46] Thus group processes, which naturally elicit justifications, don't necessarily improve on solo decision-making. And anyway, even if it could be shown that a *single* machine's decisions were less transparent than those made by a group of people, this would seem less a shortcoming of algorithms than an asymmetry in the systems being compared. A decision made by *one* person would, for the same reason, be less transparent than a decision made by a *group* of people.

Summing Up

We've tried to expose an assumption behind many of the calls for more explainable, transparent AI systems. The assumption is that it's reasonable to impose a higher standard of transparency on AI systems than would ordinarily be imposed on human decision makers. Or perhaps the assumption is simply that human decisions are generally more transparent than algorithmic decisions because they can be inspected to a greater depth, with the hefty standards imposed on machines serving merely to level the playing field. We've suggested that both assumptions are false. At this stage, the sorts of explanations we can't obtain from AI we can't obtain from humans either. On a somewhat brighter note, the sorts of explanations we *can* (and *should*) expect from human beings may be increasingly possible to obtain from AI systems.

3 Bias

Humans are remarkable thinkers. We make hundreds of decisions a day in extraordinarily complex environments with very limited information. And despite the fact that we can't possibly calculate all the consequences of our actions, we are nonetheless surprisingly successful at navigating the modern world. How are we so successful? The simple answer is we cheat. Since the 1970s psychologists have been cataloguing the many rough and ready rules that our brains apply behind the scenes to help us get important decisions right most of the time in circumstances we often meet. Known as "heuristics and biases,"[1] these psychological mechanisms help us make all the decisions we don't really have sufficient information to make. There are few types of human reasoning that don't benefit from this sort of "fast and frugal" thinking.[2] Here are some examples.

The *availability* heuristic tells us that the chance of some phenomenon happening is to be estimated by adding up how often we run into it.[3] It works well if we're trying to estimate the chance of meeting a cat, but not if we want to know, right now, the chance of being murdered. That is because media organizations make a point of informing us about everyone who gets murdered, but they don't similarly inform us about each and every cat sighting, so the availability of information about murders is deceptively high.

Object permanence describes our brains' assumption that objects continue to exist even if we're not experiencing them. It's this built-in Occam's razor that tells you that the flu you woke up with this morning is the same one you went to bed with last night. It's a good principle of critical thinking, but, like all simple "hard-wired" rules, it will inevitably occasionally lead you astray, as when the rabbit the magician pulls out of a hat isn't the one they showed you thirty seconds ago.

When we make estimates of value (How much should I pay for that car?), the *anchoring* heuristic makes us rely too heavily on an initial estimate someone else has given us. We do this even when we don't know how the estimate was reached or indeed whether it's a genuine estimate at all.[4]

In judging whether someone will, for example, be a bad driver, it would seem sensible to start by thinking about the sorts of things that cause bad driving, such as inexperience, inattention, or intoxication. But that's not the way our brains typically work. Rather than thinking about causes, we estimate how similar this person is to the central-casting stereotype of a bad driver (a bias known as *representativeness*). We're afraid of sharks because of their fearsome reputation, not the actual likelihood of encountering one, let alone being harmed by one. This so-called generic reasoning is driven by a mix of the representativeness and availability biases.[5]

The idea that we're better at making decisions than we really are is a bias known as *overconfidence*.[6] The extent and exact nature of all these biases is the subject of ongoing debate, but everyone understands their root cause. Human beings are limited reasoners. We have fallible memories. We can't take account of large amounts of data, and we can't weigh up the effects of many different factors at once. Nowhere are these failings more apparent than when *usually* reliable heuristics and biases result in decisions that are prejudiced.

Prejudices are biased decisions in favor of or against particular things, people, or groups, but not all biased decisions are prejudiced. Refusing to allow young children to drive is a bias, but it's not a prejudice. So, what's the extra ingredient that turns a bias into a prejudice? Academics disagree about exactly what prejudice is. Some think it's irrationality caused by making poor estimates of probabilities.[7] Everyone you've met from the neighboring village seemed rude, so you conclude they are all rude even though you've only met a small proportion of the village. Others think prejudice is a moral failing caused by negligence in the way we reason about groups, particularly when our prejudices justify behavior that benefits us.[8] A third explanation sees prejudice as a side effect of the sort of nonprobabilistic generic reasoning that's built into our psychological make-up and that serves us perfectly well when we avoid scary-looking dogs and food that seems "off."[9] This latter view seems particularly compelling, as it predicts correctly that much prejudice will be unconscious and stubbornly resistant to counterevidence.

What all these views have in common is that prejudice and the discrimination it yields is caused by the human inability to reason objectively from limited information about the complex environments in which we live. The problem is made worse by the powerful effects of our emotions on the objectivity of our decision-making.[10] Negative emotions like fear make us particularly prone to prejudice. Much of humanity's success is due to the development of ideas and institutions that compensate for these cognitive shortcomings: philosophical and historical analysis, the development of law and legal institutions, the scientific method, modern mathematical, statistical and logical principles, and so on. For much of our existence, these generally more accurate and objective types of reasoning have been limited in scope and applied primarily by experts, but now they're being harnessed in widely available and easily usable devices that promise objective and accurate decision-making for everyone.

Artificial Intelligence to the Rescue

Perhaps the first truly successful artificial intelligences were expert systems. These were designed to emulate the thinking of experts in specific domains, such as medical diagnostics. The subject of a great deal of work in the 1970s and 80s, expert systems never really took off for a number of reasons. They were extremely expensive to produce, and (most importantly) they could only emulate reasoning to the extent that it followed strictly definable rules. The great majority of human reasoning isn't rule-based in this strict, deterministic way. Instead, it rests on probabilities and likelihoods, risks and rewards. All the heuristics and biases we mentioned earlier evolved to help us reason about probability and value. Indeed, it's this sort of reasoning that's made humans such a successful species, so it should come as no surprise that the commercial success of AI has resulted from the development of systems capable of learning about probabilities.

The reason modern AI tools are thought to hold so much promise in minimizing bias is precisely because their power, speed, and accuracy mean they can dispense with the use of fast and frugal rules of thumb. AI can analyze large datasets. It can make decisions based on many more types of factors than humans can take into account. It isn't prone to errors in statistical reasoning of the kind humans routinely make and it doesn't use generic reasoning about stereotypes. Above all, it's relentlessly probabilistic.

It's surprising, then, that the most persistent objections to the use of AI in government, commerce, and daily life include allegations of unfairness and bias. Here are the main ones:

- it perpetuates or exacerbates existing inequalities;[11]
- it discriminates against minorities;[12]
- it hyper-scrutinizes the poor and disadvantaged;[13]
- its risk assessments in domains like justice and policing are fundamentally unfair;[14]
- it is pseudo-objective;[15]
- it evades existing protections against discriminatory reasoning based on race, gender, and other protected categories;[16]
- it obscures the often complex decisions made by developers about how to interpret and categories facts about people's lives;[17] and
- it fundamentally distorts the nature of commerce,[18] politics,[19] and everyday life.[20]

In this chapter, we'll explain and assess these allegations. Some, we will argue, are more pernicious than others. And some, due to faulty applications of probabilistic reasoning, are clearly fixable.

Taking Humans out of the Equation

Human decision-making is accurate enough in everyday circumstances, but for the reasons just discussed, the intuitive judgments of individuals aren't sufficiently reliable in high-stakes environments (Who gets a heart bypass? Who gets imprisoned?). Also, humans have a problem with partiality. Our beliefs are famously infected by our desires (colloquially known as *wishful thinking*), and evolution seems to have implanted within us a strong drive to put our own interests first, followed by those of close relatives,[21] and then those of our social groups.[22] We get around this problem in various ways in high-stakes circumstances. We vote as jurors and board members as a way of averaging out individual preferences. Social workers utilize structured decision-making when they use checkbox forms to assess risks to individuals. In many contexts, we require expert decision makers, such as judges, to set out their reasoning. But all these mechanisms are cumbersome and prone to error and deception. As we saw in chapter 2, humans are adept at rationalizing their decisions in ways that make their reasoning look nobler and more sensible than it often is. Being forewarned about the dangers of unconscious bias is surprisingly ineffective at enhancing our objectivity. Even professionals using structured decision-making tools, like checkbox forms, are known to cheat when filling out such forms in order to achieve the result that their intuitions tell them is the right one.[23] So part of the promise of AI is that in such high-stakes circumstances it could take people out of the equation, replacing them with algorithms that are reliable and impartial. But removing human bias from AI isn't as simple as it sounds. Machine learning is extraordinarily powerful, but it's humans who decide how to build and train such systems, and both the building and the training processes are open to bias. What we called "process" transparency in chapter 2 is therefore of the utmost importance.

Computer scientists are fond of pointing out that bias is not inherently a bad thing. It's part of the way we think and a very important part of what makes modern machine-learning systems so successful. It's the very fact that people's biases in what they prefer to read are reflected in the statistical patterns of their previous purchases that makes Amazon reasonably successful at predicting what they'd like to read next. Exploiting such biases allows us to develop algorithms for social sorting, market segmentation, personalization, recommendations, the management of traffic flows, and much more. But there's much to decide in working out how to successfully exploit biases in the data.

In unsupervised machine learning, developers must decide what problems they want to solve. If we're developing a system for assessing job applicants, we need to understand and prioritize the human resources challenges facing the company that will use the system. Is there a widespread mismatch between skill sets and job descriptions? Is staff turnover too high? Does the workplace need more diversity? Having settled on a problem or problems to solve, developers identify the patterns detected in the dataset relevant to those problems. Whether we're successful will ultimately depend on which data we use to train the system. How are the data collected and organized? Is the dataset diverse enough to give us accurate information about the variety of people who'll use the algorithm or on whom it'll be used? Can a system devised to detect crime in Chicago be successfully deployed in New York? What about Mumbai?

In supervised machine learning, we train the algorithm by telling it when it has the right answer. To do so, we must start by identifying success—simple enough if we're training a facial recognition system, but what if we're developing a dating app? To judge success, we'd need to understand what users wanted from a dating app. Are people looking for a life partner? If so, is there an exploitable correlation between the information people provide to dating apps and the longevity of relationships? No doubt there'll be significant disagreement about which dates are "good" from this perspective, because different people look for different things from life partners. But maybe there's *some* correlation in the data. If not, should we set our predictive sights lower—maybe the predictors of repeat dates instead of life partnerships—and hence decide not to serve the interests of those looking for long-term relationships? How much will the owners of the tool tell users about whose interests they decide to serve?

All these human choices about how algorithms are developed are further influenced by the choices of users regarding how they interact with AI and how they interpret what it tells them. Although AI continues to get better at prediction and detection tasks, researchers and journalists have identified a surprising variety of ways in which it can perpetuate disadvantage and cause harm to minorities and individuals who are far away from the population average (known by data scientists as "statistical outliers").

Predicting the Future by Aggregating the Past

"It's tough to make predictions, especially about the future." So said baseball great Yogi Berra, and he was right. We can only make accurate predictions about the future by aggregating what we know of the past. That information is inevitably incomplete and of variable quality. When we try to emulate human predictions in artificial intelligence, it often means that the algorithms we produce are fueled by aggregating the same intuitive and sometimes prejudiced human decision-making we're trying to improve on. Aggregating human decisions can be successful when we rely on the "wisdom of the crowd," but crowds aren't always wise. The wisdom of the crowd is most accurate when a diverse group of people are prone to make random errors and averaging out their judgments can effectively minimize those errors. This was famously demonstrated in 1908 by Frances Galton, who averaged out the guesses of all the contestants trying to guess the weight of a prize ox at the Plymouth County Fair. But for this to work, the guesses have to be independent of each other—members of a "wise" crowd can't influence other members.[24] You don't get an accurate estimate if guessers are free to copy the guesses of those they think are especially smart or knowledgeable. The wisdom of the crowd also fails when the errors of individuals aren't random.[25] We might all be guessing independently of one another and yet be prone to a shared bias. If, for example, humans typically estimate dark-colored animals to be heavier than light-colored ones, averaging out the guesses at the County Fair will just give us a biased average.

These two facts about the way the wisdom of crowds fails tell us something important about the dangers of developing predictive risk models fueled by data that consist of human intuitions and estimates. Normal human decision-making is replete with cognitive biases that sometimes result in systematic prejudice. This is exactly when crowds fail to be wise. So aggregating the intuitions of large numbers of individuals can produce algorithms that mirror the systematic human biases we're trying to avoid. Of course, machine learning systems are doing something much more complex than averaging. Nevertheless, fed biased data, they'll produce biased results. Nowhere are these risks more apparent than in the use of predictive risk models in policing.

Predictive policing employs artificial intelligence to identify likely targets for police intervention, to prevent crime, and to solve past crimes. Its most famous incarnation is PredPol, which began in 2006 as a collaboration between the Los Angeles Police Department and a group of criminologists and mathematicians at the University of California, San Diego. The company was incorporated in 2012 and is now used by more than sixty police departments around the US,[26] and also in the United Kingdom. PredPol makes predictions about the locations of future crime that help cash-strapped police forces allocate their resources. The predictions appear in real time as high-risk crime "hot-spots" displayed as red boxes in a Google Maps window. Each box covers an area of 150 square meters.

This is a growing industry, with PredPol now facing competition from companies such as Compustat and Hunchlab, who are racing to incorporate any information they think will help police do more with less. Hunchlab now includes Risk Terrain Analysis that incorporates features such as ATMs and convenience stores known to be locations for small-scale crime.[27] On the face of it, this seems like a sensible and admirably evidence-driven approach to crime prevention. It doesn't focus on individuals, and it doesn't know about ethnicity. It just knows crime stats and widely accepted criminological results (e.g., the likelihood of your house being burgled is strongly influenced by the occurrence of recent burglaries nearby).

But the apparent objectivity of these tools is deceptive. They suffer from the same problem that has always plagued policing. Police must make judgments about how best to prevent crime, but they only have patchy information about the incidence of crime, relying as they do on statistics about reports of crime, arrests, and convictions. Many types of crime are, by their nature, difficult to detect. You certainly know if you've been mugged, but you may well not know if you've been defrauded. Other types of crime, such as domestic violence, are persistently underreported. And of course, many reported crimes don't lead to arrests and subsequent convictions. This means that there's often a danger that the crime data informing predictive policing tools is influenced by the intuitive judgments of individual officers about where to go, whom to talk to, which leads to follow up on, and so on. It would be a Herculean task to assess the objectivity of these intuitive judgments. So, for all that predictive policing sounds scientific and the crime maps produced by companies like PredPol look objective and data-driven, we really have to accept that we cannot tell how objective the output of such tools really is. It's perhaps not surprising that civil rights groups have not been convinced by PredPol's claims that its use of only three data points (crime type, crime location, and crime date/time) eliminates the possibility for privacy or civil rights violations. A joint statement by the American Civil Liberties Union and fourteen other civil rights and related organizations focused directly on this problem:

> Predictive policing tools threaten to provide a misleading and undeserved imprimatur of impartiality for an institution that desperately needs fundamental change. Systems that are engineered to support the status quo have no place in American policing. The data driving predictive enforcement activities—such as the location and timing of previously reported crimes, or patterns of community- and officer-initiated 911 calls—is profoundly limited and biased.[28]

It's tempting to think that this problem of tools relying on data contaminated by human intuitions will gradually dissipate over time as police come to rely more on statistically accurate algorithms and less on the intuitions of officers, but the actual effect of developing a predictive risk model based on a systematically biased dataset can be to bake in bias rather than to allow it to gradually dissipate. Kristian Lum and William Isaac show that predictive policing models are likely to predict crime in areas already believed to be crime hotspots by police.[29] These areas will then be subject to more police scrutiny, leading to observation of criminal behavior that confirms the prior beliefs of officers about where crime is most common. These newly observed crimes are fed back into the algorithm, generating increasingly biased predictions: "This creates a feedback loop where the model becomes increasingly confident that the locations most likely to experience further criminal activity are exactly the locations they had previously believed to be high in crime: selection bias meets confirmation bias."[30]

These problems aren't inevitable and we certainly don't mean to suggest that predictive policing strategies are useless. We want high-stakes decisions to be data-driven and predictive risk models are likely to be more accurate and more reliable at making predictions from those data. The problem, as always in policing, is getting better data. Essentially the same problem strikes algorithms designed to help make decisions about the allocation of bank loans, medical procedures, citizenship, college admissions, jobs, and much else besides. So it's a good sign that significant effort is being put into tools and techniques for detecting bias in existing datasets. As of 2019, major tech companies including Google, Facebook, and Microsoft, have all announced their intention to develop tools for bias detection, although it's notable that these are all "in-house." External audits of biases in the artificial intelligence developed by those companies would probably be more effective. That said, bias detection is only the start, particularly in domains like policing. Many countries already know that their arrest and incarceration statistics are heavily biased toward ethnic minorities. The big question is, can we redesign predictive policing models, including rules about how they're used and how their source data are collected, so that we're at least not further disadvantaging poor "crime-ridden" communities?

At this point, the most important thing governments and citizens can do is to acknowledge and publicize the dangers of the algorithms that suffer from the "bias in/bias out" problem. It'll then be in the interests of the makers and owners of such algorithms to modify their products to help minimize this type of bias. It must become common knowledge that AI predictions are only as good as the data used to drive them. The onus is at least partly on us to be savvy consumers and voters.

To better understand how to tackle algorithmic bias, we're going to need to know more about the ways human biases can creep into the development of AI. There are many ways to categorize algorithmic biases, but we'll divide them into three groups: biases in how we *make* AI; biases in how we *train* it; and biases in how we *use* it in particular contexts.

Built-in Bias

Human bias is a mix of hardwired and learned biases, some of which are sensible (such as "you should wash your hands before eating"), and others of which are plainly false (such as "atheists have no morals"). Artificial intelligence likewise suffers from both built-in and learned biases, but the mechanisms that produce AI's built-in biases are different from the evolutionary ones that produce the psychological heuristics and biases of human reasoners.

One group of mechanisms stems from decisions about how practical problems are to be solved in AI. These decisions often incorporate programmers' sometimes-biased expectations about how the world works. Imagine you've been tasked with designing a machine learning system for landlords who want to find good tenants. It's a perfectly sensible question to ask, but where should you go looking for the data that will answer it? There are many variables you might choose to use in training your system—age, income, sex, current postcode, high school attended, solvency, character, alcohol consumption? Leaving aside variables that are often misreported (like alcohol consumption) or legally prohibited as discriminatory grounds of reasoning (like sex or age), the choices you make are likely to depend at least to some degree on your own beliefs about which things influence the behavior of tenants. Such beliefs will produce bias in the algorithm's output, particularly if developers omit variables which are actually predictive of being a good tenant, and so harm individuals who would otherwise make good tenants but won't be identified as such.

The same problem will appear again when decisions have to be made about the way data is to be collected and labeled. These decisions often won't be visible to the people using the algorithms. Some of the information will be deemed commercially sensitive. Some will just be forgotten. The failure to document potential sources of bias can be particularly problematic when an AI designed for one purpose gets coopted in the service of another—as when a credit score is used to asses someone's suitability as an employee. The danger inherent in adapting AI from one context to another has recently been dubbed the "portability trap."[31] It's a trap because it has the potential to degrade both the accuracy and fairness of the repurposed algorithms.

Consider also a system like TurnItIn. It's one of many anti-plagiarism systems used by universities. Its makers say that it trawls 9.5 billion web pages (including common research sources such as online course notes and reference works like Wikipedia). It also maintains a database of essays previously submitted through TurnItIn that, according to its marketing material, grows by more than fifty thousand essays per day. Student-submitted essays are then compared with this information to detect plagiarism. Of course, there will always be some similarities if a student's work is compared to the essays of large numbers of other students writing on common academic topics. To get around this problem, its makers chose to compare relatively long strings of characters. Lucas Introna, a professor of organization, technology and ethics at Lancaster University, claims that TurnItIn is biased.[32]

TurnItIn is designed to detect copying but all essays contain something like copying. Paraphrasing is the process of putting other people's ideas into your own words, demonstrating to the marker that you understand the ideas in question. It turns out that there's a difference in the paraphrasing of native and nonnative speakers of a language. People learning a new language write using familiar and sometimes lengthy fragments of text to ensure they're getting the vocabulary and structure of expressions correct.[33] This means that the paraphrasing of nonnative speakers of a language will often contain longer fragments of the original. Both groups are paraphrasing, not cheating, but the nonnative speakers get persistently higher plagiarism scores. So a system designed in part to minimize biases from professors unconsciously influenced by gender and ethnicity seems to inadvertently produce a new form of bias because of the way it handles data.

There's also a long history of built-in biases deliberately designed for commercial gain. One of the greatest successes in the history of AI is the development of recommender systems that can quickly and efficiently find consumers the cheapest hotel, the most direct flight, or the books and music that best suit their tastes. The design of these algorithms has become extremely important to merchants—and not just online merchants. If the design of such a system meant your restaurant never came up in a search, your business would definitely take a hit. The problem gets worse the more recommender systems become entrenched and effectively compulsory in certain industries. It can set up a dangerous conflict of interest if the same company that owns the recommender system also owns some of the products or services it's recommending.

This problem was first documented in the 1960s after the launch of the SABRE airline reservation and scheduling system jointly developed by IBM and American Airlines.[34] It was a huge advance over call center operators armed with seating charts and drawing pins, but it soon became apparent that users wanted a system that could compare the services offered by a range of airlines. A descendent of the resulting recommender engine is still in use, driving services such as Expedia and Travelocity. It wasn't lost on American Airlines that their new system was, in effect, advertising the wares of their competitors. So they set about investigating ways in which search results could be presented so that users would more often select American Airlines. So although the system would be driven by information from many airlines, it would systematically bias the purchasing habits of users toward American Airlines. Staff called this strategy *screen science*.[35]

American Airline's screen science didn't go unnoticed. Travel agents soon spotted that SABRE's top recommendation was often worse than those further down the page. Eventually the president of American Airlines, Robert L. Crandall, was called to testify before Congress. Astonishingly, Crandall was completely unrepentant, testifying that "the preferential display of our flights, and the corresponding increase in our market share, is the competitive raison d'être for having created the [SABRE] system in the first place."[36] Crandall's justification has been christened "Crandall's complaint," namely, "Why would you build and operate an expensive algorithm if you can't bias it in your favor?"

Looking back, Crandall's complaint seems rather quaint. There are many ways recommender engines can be monetized. They don't need to produce biased results in order to be financially viable. That said, screen science hasn't gone away. There continue to be allegations that recommender engines are biased toward the products of their makers. Ben Edelman collated all the studies in which Google was found to promote its own products via prominent placements in such results. These include Google Blog Search, Google Book Search, Google Flight Search, Google Health, Google Hotel Finder, Google Images, Google Maps, Google News, Google Places, Google+, Google Scholar, Google Shopping, and Google Video.[37]

Deliberate bias doesn't only influence what you are offered by recommender engines. It can also influence what you're charged for the services recommended to you. Search personalization has made it easier for companies to engage in *dynamic pricing*. In 2012, an investigation by the Wall Street Journal found that the recommender system employed by a travel company called Orbiz appeared to be recommending more expensive accommodation to Mac users than to Windows users.[38]

Learning Falsehoods

In 2016, Microsoft launched an artificially intelligent chatbot that could converse with Twitter users. The bot called "Tay" (which stands for "thinking about you") was designed to mimic the language patterns of a nineteen-year-old American girl. It was a sophisticated learning algorithm capable of humor and of seeming to have beliefs about people and ideas.[39] Initially, the experiment went well, but after a few hours, Tay's tweets became increasingly offensive. Some of this invective was aimed at individuals, including President Obama, consisting of false and inflammatory claims about events and ethnicities. In the sixteen hours and ninety-three thousand tweets before Microsoft shut it down, Tay had called for a race war, defended Hitler, and claimed that Jews had caused the 9/11 attacks.[40]

Microsoft's explanation of what went wrong was limited to "a coordinated attack by a subset of people exploited a vulnerability in Tay. Although we had prepared for many types of abuses of the system, we had made a critical oversight for this specific attack."[41] There is little doubt that Tay was attacked by trolls who were deliberately feeding it false and offensive information, but perhaps the more important problem is that Tay wasn't in a position to know that the trolls *were* trolls. It was designed to learn, but the vulnerability to which Microsoft refers left it without the capacity to assess the quality of the information it was given. Certainly it lacked the cognitive complexity, social environment, and educational background of a real teenager. In short, Tay was a learner but not a *thinker*.

As we mentioned in the prologue, we may one day invent a general AI, capable of understanding what it says and possessing a level of complexity that would allow it to assess the truth of the data it received, but we're not there yet, and may not be for a long time to come. Until then we're stuck with narrow AI that is completely dependent on human beings supplying it with accurate and relevant data. There are several ways in which this can go awry.

Clearly Tay wasn't the only one being lied to on the internet in 2016. From people lying on dating sites to lying about hotels on TripAdvisor, there's no shortage of individuals happy to submit false information into recommender engines. Although progress is being made in using machine learning to detect lying online, at least the more savvy users of sites relying on aggregated user contributions retain some skepticism about the views they express.

More concerning are cases in which a training set contains data that accurately represent one part of a population, but don't represent the population as a whole—the phenomenon we called "selection" (or "sampling") bias earlier. In a famous recent instance of this problem, Asian users of Nikon cameras complained that the camera's software was incorrectly suggesting that they were blinking.[42] Users have similarly complained that tools relying on speech recognition technology, such as Amazon's Alexa, are inaccurate at interpreting some accents. They're particularly poor at recognizing the speech of Hispanic and Chinese Americans.[43]

These issues can have serious consequences in high-stakes environments. The feeds of CCTV cameras are now routinely assessed using facial recognition systems, which have been shown to be particularly sensitive to the racial and gender diversity of the data on which they're trained.[44] When these algorithms are trained on criminal databases for use in law enforcement, it's not surprising that they show biases in their abilities to recognize the faces of different groups of people. They identify men more frequently than women; older people more frequently than younger; and Asians, African Americans, and other races more often than whites.[45] As a result, false accusations or unnecessary questioning is more likely to land on women and members of racial minorities.

Are we being too hard on the makers of facial recognition algorithms? After all, something like this problem occurs when humans recognize faces. The "other-race" effect refers to the fact that humans are better at recognizing the faces of people of their own ethnicity.[46] So the problem isn't that AI makes mistakes—people make mistakes too. The problem is that the people using these systems might be inclined to think that they're bias-free. And even if they're aware of issues like selection bias, users may still not be in the best position to identify the strength and direction of bias in the tools they use.

But the news isn't all bad. Well trained, these systems can be even better than humans. There is much we can do to increase the diversity and representativeness of the databases on which such algorithms are trained. And although there are barriers to the collection and processing of quality data owing to anti-discrimination and intellectual property law (e.g., copyright in images), they aren't insurmountable.[47] Such problems can also be addressed when people purchasing or using such systems make a point of asking what the developers have done to ensure that they are fair.

Bias in the Context of Use

A 2015 study detected that Google's ad-targeting system was displaying higher-paying jobs to male applicants on job search websites.[48] It may be that this algorithm had built-in biases or that something could've been done to address selection bias in the dataset on which it was trained. But putting those issues aside, there is likely something else going on—something at once more subtle but also (in a way) obvious. All countries suffer from gender pay gaps. What the system may have *correctly* been detecting is the fact that, on average, women are paid less than men. What it couldn't detect is that that fact about the world is unjust. The fact that women don't do as much high-paid work has nothing to do with their inability to perform such tasks. Rather, it reflects a history of gender stereotypes, discrimination, and structural inequality that society is only now beginning to address systematically. Such cases are instructive for those who develop and deploy predictive algorithms. They need to be attuned to the ways in which a "neutral" technology, applied in particular circumstances, can perpetuate and even legitimize existing patterns of injustice. As a society changes, its use of algorithms needs to be sensitive to those changes.

Sadly, problems like these are difficult to address. We can, of course, remove variables such as gender and ethnicity from the datasets on which we train such algorithms, but it's much more difficult to remove the *effect* of such factors in those datasets. The makers of PredPol deliberately avoided using ethnicity as a variable in their predictions about the locations of future crime, but PredPol is all about geography, and geography is strongly correlated with ethnicity. Data scientists would say that geography is a *proxy variable* for ethnicity, so the result of using such systems is often to focus police scrutiny on minority communities.

Can AI Be Fair?

COMPAS (Correctional Offender Management Profiling for Alternative Sanctions) is described by its makers, the company Northpointe (now Equivant), as a "risk and needs assessment instrument," used by criminal justice agencies across the US "to inform decisions regarding the placement, supervision, and case management of offenders."[49] It consists of two primary predictive risk models designed to predict the likelihood of recidivism amongst prisoners in general and specifically amongst violent offenders. The details of its design are a trade secret but prisoners encounter it as a lengthy questionnaire that is used to generate a risk score between one and ten. That score is used in many contexts within the justice system, including decisions about where they're imprisoned and when they're released.

In 2016, ProPublica, an independent journalism organization, conducted a two-year study of more than ten thousand criminal defendants in Broward County, Florida. It compared their predicted recidivism rates with the rate that actually occurred. The study found that "black defendants were far more likely than white defendants to be incorrectly judged to be at higher risk of recidivism, while white defendants were more likely than black defendants to be incorrectly tagged as low risk."[50]

Northpointe strongly denied the allegation, claiming that ProPublica made various technical errors (which ProPublica denied). Northpointe also argued that ProPublica had not taken into account the fact that recidivism is higher amongst the black prison population. ProPublica was unconvinced. Black Americans are more likely to be poor which diminishes their educational opportunities. They're also more likely to live in high-crime neighborhoods with fewer job opportunities. If these facts are even part of the explanation for their higher recidivism rates, it seems unjust that they be further penalized by an algorithm that decreases their chance of parole. It soon became clear that the debate between Northpointe and ProPublica was fundamentally a disagreement about how an algorithm such as COMPAS could be made fair.

It's true that we can develop algorithms that are, in some sense, "fairer." The challenge, however, is that there are many conflicting algorithmic interpretations of fairness, and it isn't possible to satisfy all of them. Three common definitions of fairness are discussed in the machine learning community.

- We can ensure that protected attributes, such as ethnicity and gender (and proxies for those attributes), aren't explicitly used to make decisions. This is known as *anti-classification.*
- We can ensure that common measures of predictive performance (e.g., false positive and false negative rates) are equal across groups defined by the protected attributes. This is known as *classification parity.*
- We can ensure that, irrespective of protected attributes, a risk estimate means the same thing. A high recidivism risk score, for example, should indicate the same likelihood of reoffending regardless of the ethnicity, gender, or other protected attribute of the assessed person. This is known as *calibration.*

It's not hard to show using a bit of high-school algebra that when the incidence (or "base rate") of a phenomenon like recidivism differs across distinct populations (e.g., ethnic groups), you simply can't simultaneously satisfy all such criteria; some of them are mutually exclusive.[51] So Although Northpointe could defend COMPAS by showing it to be well calibrated (risk scores mean the same regardless of group membership),[52] ProPublica was nonetheless able to claim that COMPAS is biased because it doesn't satisfy classification parity (specifically, *error rate balance*—where the rate of false positives and false negatives is equal across groups). As we saw, ProPublica found that black people assessed with COMPAS were more likely than white people to be incorrectly classified as high risk (so the rate of false positives differed across the two groups), whereas white people were more likely than black people to be incorrectly classified as low risk (so the rate of false negatives differed across the two groups). Fact is, you can't satisfy both calibration and error rate balance if the base rates differ—and, of course, for a variety of reasons (including historical patterns of prejudicial policing), recidivism base rates *do* differ between black and white populations in the United States.

There are also trade-offs between fairness and accuracy. For example, in the COMPAS case, anti-classification would require that the data used didn't contain any proxies for race. As noted above, race is strongly correlated with many important facts about people including income, educational achievement, and geographic location. So removing such variables would be likely to make the algorithm much less accurate and therefore much less useful.

Recent work in data science shows that conflict between these various notions of fairness isn't limited to the COMPAS case and isn't always a result of design failures by those developing such algorithms. These different characterizations of fairness are logically incompatible. As scholars have noted, "there is a mathematical limit to how fair any algorithm—or human decision maker—can ever be."[53] Nor are these conflicts purely technical, either: they're as much political. Choosing between different standards of fairness calls for informed discussion and open, democratic debate.

Summing Up

This chapter might seem unfair to those developing AI, listing as it does a surprising variety of ways in which AI can be biased, unfair, and harmful, but we're not Luddites, and don't mean to portray AI itself as the problem. Many of the issues raised in this chapter affect human reasoning just as much as they affect AI. The incompatibility of anti-classification, classification parity, and calibration is first and foremost a fact about *fairness*, not AI. Human intuitions about fair decision-making are at odds with each other, and this tension is reflected in the statistical rules that formalize those intuitions. Likewise, algorithmic bias is parasitic on human bias, and, if anything, the effect of human bias is in some ways worse than algorithmic bias. The sort of generic reasoning that leads to human prejudice isn't only harmful; it's irrational, often unconscious, and insensitive to counter-evidence. By contrast, many of the biases from which AI suffers can be detected and, at least in principle, remedied. That said, there are few easy fixes. Algorithms can't be completely fair, and some ways of enhancing their fairness (such as anti-classification) make them less accurate.

So, the ball is in our court. As consumers and voters, we'll need to thrash out the sort of fairness we want in the many domains in which AI is used. This, of course, means we'll have to decide how much *unfairness* we're prepared to put up with.

4 Responsibility and Liability

One of the greatest sources of angst around the advent of increasingly intelligent computer systems is the alleged erosion of human responsibility they'll entail. International discussions on the use of lethal autonomous weapons systems provide a forceful illustration of what's at stake. If weapons systems powered by complex and potentially opaque machine-learning technology will soon be able to decide, unassisted, when to engage a target, will we still meaningfully be able to hold human beings responsible for war crimes involving these systems? In war we're obviously talking about life and death decisions, but similar worries play out elsewhere—areas in which lives may not be directly on the line but in which advanced systems can nevertheless have significant effects on people and society, such as in law enforcement, transportation, healthcare, financial services, and social media. If AI technologies can decide things without direct human intervention in ways that are too complex or too fast for human beings to properly understand, anticipate, or change, will we still be able to hold human beings responsible for these systems? And would we *want* to hold them responsible? What would be the alternative? Can *machines* be responsible?

What exactly *is* responsibility, and what qualities of AI challenge existing ways of distributing it? Does responsibility simply vanish? Does it fade away gradually, or is something else happening entirely? In this chapter we'll look at some of these questions. We'll begin by unpacking some of the different senses of responsibility, including moral and legal responsibility. We'll then highlight some of the ways in which technology, in general, affects these senses. In the remainder of the chapter we'll home in on some of the discussions about various aspects of responsibility that developments in AI—as a special kind of technology—have generated.

Picking Apart Responsibility

Responsibility is one of those concepts most of us *think* we understand. Within moral philosophy, legal scholarship, and even daily conversation, the word is often used interchangeably with words like "accountability," "liability," "blameworthiness," "obligation," and "duty."[1] Yet there are nuanced differences between these terms. Consider this little snippet from the philosopher H. L. A. Hart, describing a (fictional) sea captain:

> As captain of the ship, [he] was responsible for the safety of his passengers and crew. But on his last voyage he got drunk every night and was responsible for the loss of the ship with all aboard. It was rumored that he was insane, but the doctors considered that he was responsible for his actions. Throughout the voyage he behaved quite irresponsibly, and various incidents in his career showed that he was not a responsible person. He always maintained that the exceptional winter storms were responsible for the loss of the ship, but in the legal proceedings brought against him he was found criminally responsible for his negligent conduct, and in separate civil proceedings he was held legally responsible for the loss of life and property. He is still alive and he is morally responsible for the deaths of many women and children.[2]

Hart's story isn't supposed to be riveting—which is just as well!—but it does make its point vividly: responsibility is a complex, multifaceted notion. It can refer to causal contribution, as when winter storms cause the loss of a ship or to the formal mechanism used to attribute legal responsibility. It can be used to describe a character trait and also the obligations and duties that come with the professional role of a sea captain. It can mean any or all of these things and other things besides. Responsibilities are also by nature diffuse. Events can rarely be pinned down to one person or factor alone. Why was the drunken captain allowed to be captain if he had a history of alcoholism? Why wasn't this picked up? Who cleared the captain as safe? Who designed the screening protocol?

In this chapter, we're not going to consider all the forms of responsibility Hart explored in his story. Our focus instead will be on moral and legal notions of responsibility, since it's at these points where the rubber hits the road, so to speak, as far as artificial intelligence is concerned. Let's start by considering how they differ.

First, moral responsibility is very often (but not always) *forward*-looking, and legal responsibility is very often (but not always) *backward*-looking.[3] This is a crucial point, in fact, because many of the differences between these two forms of responsibility can be traced back to this one pivotal distinction. What does it *mean* to be forward-looking or backward-looking?

Responsibility is forward-looking when it relates to the prospective duties and obligations a person may have. Expecting a certain degree of care and civility in human relationships is normal. Every time we step into a car, we take on various obligations—some to pedestrians, some to motorists, others to public and private property-holders. In our workplaces we expect that a certain level of professional courtesy will be extended to us. In the domain of goods and services, we naturally expect that manufacturers will have an eye to safe construction and assembly. And when it comes to autonomous systems, we naturally expect that software engineers will take the risks of certain design decisions into account when configuring their programs. If they know a credit-risk assessment tool has a good chance of unfairly disadvantaging a particular group of people, they have a responsibility—if not a legal one than certainly a moral one—to adjust the system in order to reduce the risk of unfair bias. Such forward-looking responsibilities are about events that will happen in the future and are the bread and butter of instruction manuals, professional codes, regulations, training programs, and company policies.

Backward-looking responsibility is different altogether; it's about accountability for events that have *already* happened. If an accident occurs, chains of responsibility will be traced back in order to determine who was at fault and should be called on to answer for their actions. Backward-looking responsibilities most often result in legal liability (see below), or some other kind of formal or informal reckoning.

So in what other ways do moral and legal responsibility differ? Well, let's take a step back for a moment.

A useful way of thinking about *any* kind of responsibility is to carefully consider the relationships in which it actually arises and how burdens of responsibility are distributed within them. After all, responsibility is fundamentally a relational concept—it's about what we owe each other and the demands it's reasonable to place on our fellows.

Relationships of responsibility have several components. In a simple breakdown,

- the *agent* is the person (or group) who performs the action;
- the *patient* is the person (or group) who receives or is somehow affected by the action;

- the *forum* is the person (or group) who determines who should be responsible (this can be an agent reflecting on their own conduct, but it can also be a judge, the general public, or some other entity);
- the *enforcement mechanism* is the means by which censure or approbation is registered (the forum assigns blame or praise and maybe even punishment based on commonly accepted standards of adjudication).

Various configurations of these components lead to differences in the kinds of responsibility we can expect in particular relationships. We'll see, for example, that the burden of legal responsibility often mischievously shifts back and forth between agent and patient, depending on the context. But let's consider moral responsibility first.

In traditional moral philosophy, moral responsibility is firmly human-centered.[4] Particularly in liberal democracies, human beings are commonly thought to possess autonomy and free will, which is the basis for being held morally responsible. This capacity is what sets us apart from machines and animals. And although human autonomy should definitely be celebrated, there *are* strings attached. Precisely because we get to decide for ourselves—assuming there is no one forcing or coercing us to do otherwise—we *must* live with the consequences of our choices. The flip side of being free to choose as we please is that in the end the buck stops with us.

Autonomy is a necessary condition for the attribution of moral responsibility (we'll unpack it a little more in chapter 7), but there are two other conditions frequently discussed in moral philosophy.[5] One is that there should be a causal connection between the conduct of an agent and the outcome of events; the other is that the agent must be able to foresee the predictable consequences of their actions. To be held morally responsible, a person should knowingly be able to change the outcome of events in the world and thus foresee (to some extent) the likely consequences of their actions. It simply wouldn't be reasonable blaming someone for harm they couldn't have known would result from their actions.

What exactly these three conditions mean and to what extent they properly ground responsibility is subject to continuing debate within moral philosophy. There are plenty of open questions here. Are we *really* free to act or are our actions determined more by nature, nurture, and culture? Do we *really* have control over the outcome of events? To what extent can we know the outcome of events? Yet a common feature of many of these debates is that the agent takes center stage—on this there is little disagreement.

Debates focus on the conditions for an agent to be reasonably held responsible, such as possession of a sound mind. The patient tends to disappear from view or is too easily dismissed as a passive, extraneous factor with no independent bearing. Similarly, the forum is an abstract and inconsequential entity—the moral philosopher, professional opinion, or perhaps the wider moral community.

Certain kinds of legal responsibility have a much broader view of the responsibility relationship. Legal responsibility, generally called "liability," is related to and may overlap with moral responsibility, but it's not quite the same thing. Someone can be legally responsible without being considered morally responsible. In many countries, the driver of a vehicle can be held liable for an accident without clear evidence of moral wrongdoing.[6] Indeed the very word "accident" itself denotes the arbitrary and happenstance character of the event—the fact that no one is truly to blame. For instance, if you're driving along a road and happen to get stung by a bee, swerve, and crash into an oncoming vehicle, it's unlikely anyone would be morally blameworthy if the sting was, let's say, to your eyelid, and unavoidable by taking all reasonable precautions. Conversely, we might consider that social media platforms have a moral responsibility to tackle the problem of fake news, but in most jurisdictions they have no legal responsibility to do so and can't be held liable for harms caused by this form of misinformation.

Liability is to some extent a contrivance designed to regulate the behavior of citizens.[7] As such, it comes in different shapes and sizes depending on the legal system and which behavior it's meant to regulate. Criminal liability, for instance, focuses primarily on the conduct and mental state of the agent (and is thus a little more like moral responsibility), whereas civil liability places relatively more emphasis on the consequences for the patient and the need to distribute the burdens of harm fairly.[8]

Even *within* the area of civil liability the emphasis can shift again—there's no way to predict where the burden will fall just by knowing that a case is a "civil" rather than "criminal" matter. Take *fault-based liability* and *strict liability*. Fault-based liability requires proof that someone did something wrong or neglected to act in a certain way. In order for a person to be held liable, there needs to be evidence of a causal link between the outcome of events and the actions or inactions of the person. We expect manufacturers to ensure that the products they sell are safe and won't cause harm. If a microwave is defective, the manufacturer can be held liable for any harm caused by the microwave if it can be shown that the manufacturer failed to take reasonable precautions to minimize the risk of harm. However, it can be difficult to establish causal connections between conduct (or omissions) and outcomes, making it hard for victims to be compensated for harm. The burden of the incident will then squarely fall on them, which many people may consider unfair.

Strict liability provides a way to even out the unfairness that fault-based liability may entail. Although strict liability comes in many forms, the general idea is that it doesn't require evidence of fault. Where fault-based liability looks at the conduct and intentions of agents and how much control over the outcome of events they have, strict liability shifts most of its attention onto the patient. For defective products that cause harm, this means that the victim only has to show that the product was defective and that the defect caused harm. The victim doesn't have to prove that the manufacturer was negligent, that is, failed to take reasonable precautions against harm—it's enough to show that the product was defective and caused the victim harm. This kind of liability makes it easier for victims to be compensated and places some of the burden on the actors in the relationship that are better placed to absorb the costs than the victims.

Some forms of strict liability don't even require the identification of individual victims—the patient can be a group, society in general, or even the environment! Think of human rights. As the legal scholar Karen Yeung points out, any interference with human rights, such as freedom of speech, attracts responsibility without proof of fault even if there are no clear victims.[9] By taking the agents, patients, and potentially the whole of society into account, as well as the fair distribution of the burdens of harm, legal responsibility has a much wider gaze than moral responsibility.

Another difference between moral and legal forms of responsibility is that moral responsibility most commonly attaches to individuals, whereas it's not at all unusual to hear of companies or conglomerates being held legally responsible for their misdeeds and liable to pay sometimes multimillion dollar damages or fines for breaches of consumer safety or environmental protection laws.

We should note one final, important difference here too. The sanction itself (e.g., fines, compensation, etc.), as well as the forum that determines the sanction (e.g., a tribunal or local council authority), are both crucially important considerations when establishing legal responsibility. It very much matters *who* is deciding the outcome and *what* the outcome is. If a company errs, the denunciations of commercial competitors won't carry much weight with the offender. The rulings of a supreme court, on the other hand, surely will. The sanction, too, is paramount in law. Remember that liability tends to be a backward-looking form of responsibility in which agents have to make amends. Although these sanctions are often imposed for breaches of forward-looking duties, the emphasis is on paying up (retribution), *giving* up (disgorgement), or giving *back* (restitution). These features of legal responsibility aren't always present in the moral sphere—the sanction in particular has little relevance, although admittedly, *who* expresses moral disapproval can matter a great deal to the moral wrongdoer. Children, for instance, may be more affected by their teachers' rebukes than by their parents'. Of course, moral disapproval can result in social consequences—ostracism, injury to reputation, and the like—but these "sanctions" aren't usually formal, planned, or structured (the way a fifty-hour community service order is). More often than not, they arise from the instinctive resentments people feel when they've been mistreated.

Technology and Responsibility

The philosopher Carl Mitcham noted that technology and responsibility seem to have co-evolved since the industrial revolution and the rise of liberal democracy.[10] Responsibility filled a gap that was created by the introduction of industrial technologies. These technologies extended human capacities, enabling human beings to do things that they couldn't do before and giving them tremendous power over nature and each other. As the levers of control got longer and the distance between action and consequence greater, discussions grew about how that increasing power could be held in check. Responsibility was a solution: with great power must come great responsibility.

Mitcham's observation underscores the special relationship between technology and responsibility. The introduction of a technology changes human activity in ways that affect the conditions for the attribution of responsibility. If we take moral responsibility as an example here, technologies affect the extent to which human beings are in control or free to act, how their actions contribute to outcomes, and their ability to anticipate the consequences of their actions. Let's take these in turn.

Freedom to Act

Technologies can affect the freedom we have to make informed choices. On the one hand, they can augment our capabilities and broaden the set of options available to us. On the other hand, they frequently constrain these very capabilities and options. The internet affords us a virtually limitless space of opinions and information we can freely indulge. At the same time—and as we'll see in chapters 6 and 7—it's the data collected about us while online that can be parlayed into the targeting algorithms that restrict the kinds of opportunities, opinions, and advertising we're exposed to.

Think also of the automated administrative systems that make it easier to process large numbers of cases efficiently (see chapter 8). These systems, *by design*, reduce the discretionary scope of lower-level bureaucrats to make decisions on a case-by-case basis.[11] Technologies can empower us, to be sure, but they can also rein us in. And the better they are at what they do, the more difficult it becomes to manage without them. They have a way of inducing dependency. Does anyone under the age of twenty-five even *know* what a street map is? Even those of us over twenty-five must admit we'd find life just a little harder without Google Maps on our phones.

Causal Contribution

Technology can obscure the causal connections between the actions a person takes and the eventual consequences. It wouldn't make sense to blame someone for something they had limited control over. Complex technological systems are problematic in this respect, because they often require the collective effort of many different people in order to work. The difficulty of finding responsible individuals among the many who contributed to the development and use of a technological system is known as *the problem of many hands*.[12] It can be a real challenge ascribing responsibility in highly technological environments when there isn't a single individual with complete control or knowledge of how events will turn out. Pilots aren't the only ones needed to keep planes aloft. An aircraft is a staggeringly complex piece of kit that incorporates many different subsystems and personnel. None of the people involved has direct control of what's happening. None has a complete understanding of all the components in the operation. Air traffic controllers, maintenance personnel, engineers, managers, and regulators all have a role to play in ensuring the safe flight of an aircraft. When an accident occurs, it's often the result of an accumulation of minor errors that on their own might not have turned out disastrously. This isn't to say that no one's responsible at all. Each actor contributed in a small way to the outcome and thus has at least *some* responsibility for what took place, but the cooperative nature of the endeavor often makes it extremely difficult to isolate individual contributions. It's not quite as difficult as isolating the bananas, berries, and kiwis in a blended concoction, but it can be *nearly* as difficult!

Adding to the problem of many hands is the temporal and physical distance that technologies can create between a person and the consequences of their actions. This distance can blur the causal connection between actions and events. Technologies extend the reach of human activity through time and space. With the help of communication technologies people can, for example, interact with others on the other side of the world. Such distances can limit or change how agents experience the consequences of actions and as a result limit the extent to which they feel responsible. Sending a mean tweet may be easier than saying it directly to someone's face because you don't directly see the consequences of your actions. Similarly, the designers of an automated decision-making system will determine ahead of time how decisions should be made, but they'll rarely see how these decisions will impact the individuals they affect—impacts that may only be felt years later.

The connection between the designers' choices and the consequences of their choices is further blurred because the design often doesn't determine exactly how the technology will be used. We, as users, often still get to decide how and when to use these technologies. We can even put them to uses that designers may not have expected. Students with a smart phone no longer need to spend hours in the library hunched over a photocopier. Why bother, when you can take snaps of the relevant pages with your phone? Did the engineers who made smart phone camera functionality a reality ever imagine it would replace a photocopier? Possibly, but it surely wouldn't have occurred to everyone or been uppermost in their minds.

Foreseeing the Consequences of One's Actions

The distancing effect that technologies can have not only enables and constrains human activities, it also mediates our relationship to the future. The various sensors and measurement instruments in a cockpit translate a range of observations such as the altitude or pitch of a plane into numbers and signs that the pilot can use to inform their awareness of the situation. In so doing, these instruments help the pilot better anticipate the consequences of particular decisions. At the same time, however, the pilot may only have a partial understanding of the mechanisms, assumptions, models, and theories behind the technology being used, and this very opacity can itself compromise a pilot's ability to assess the consequences of particular decisions.

The novelty of a technology can also affect our foresight. It requires knowledge, skill, and experience to operate a technology responsibly and appreciate how it behaves under different conditions. The learning curve is steeper for some technologies than for others. In general, inexperience definitely complicates things.

Again it's important to stress that none of this makes responsibility a mirage. We've managed to create relatively successful practices in which different, complementary forms of responsibility, including legal, moral, and professional responsibility, are distributed across multiple actors despite the problem of many hands. Some of these forms of responsibility don't require direct or full control nor even the ability to foresee likely consequences, and when technologies constrain the autonomy of one individual in a chain, responsibility is often redistributed to other actors higher up in the chain. After all, the effects of technology on human conduct are still a function of the activities of the human agents that created them and made them available. People create and deploy technologies with the objective of producing some effect in the world. The intentions of developers to influence users in particular ways are inscribed within the technology. This kind of control and power necessarily attracts responsibility—although whether this continues to be the case depends on the kind of agency technology can be expected to acquire in the future.

AI and Responsibility

As we've seen, concerns about responsibility in light of technological developments aren't new, yet AI technologies seem to present new challenges. Increasing complexity, the ability to learn from experience, and the seemingly autonomous nature of these technologies suggest that they are qualitatively different from other computer technologies. If these technologies are increasingly able to operate without the direct control or intervention of human beings, it may become hard for software developers, operators, or users to grasp and anticipate their behavior and intervene when necessary. Some suggest that, as these technologies become more complex and autonomous, we can't reasonably hold human beings responsible when things go wrong. The philosopher Andreas Matthias calls this the "responsibility gap"—the more autonomous technologies become, the less we can hold human beings responsible.[13] Some therefore argue that there may come a point at which we should hold AI technologies responsible and attribute them some kind of legal personhood or moral agency.[14] However, before addressing this suggestion, let's take a closer look at the alleged "responsibility gap."

Underlying the idea of a responsibility gap are several assumptions about responsibility. One assumption is that human beings must have direct control over the outcome of their actions before they can be held responsible for them. However, as we've seen, this narrow notion of responsibility doesn't accurately reflect how we deal with responsibility in many cases, since various notions of responsibility make do without direct control.

Another set of assumptions underlying the idea of a "responsibility gap" has to do with AI itself. AI in these debates is often unhelpfully framed as an independent, monolithic thing possessing certain human-like abilities, but this is misleading in several ways. First, it misses the larger context in which the technology is embedded and the human beings "standing just off stage."[15] In truth, it requires considerable work from human beings for current AI technologies to operate independently. Not only do humans have to design, develop, deploy, and operate these systems, they also have to adjust themselves and their environments to make sure that the technologies work successfully. Self-driving cars aren't independent of the roads they drive on or the interests of the cyclists and pedestrians they come across. These cars operate in environments regulated by rules intended to minimize risks and balance the competing interests of the various actors present in those environments. In short, for an autonomous technology to work, many human beings have to make many decisions, and these all involve choices for which they could be held responsible.

It's also clear that underlying worries about a "responsibility gap" is the assumption that machine autonomy and human autonomy are essentially similar. But there are significant differences between the two. Human autonomy is a complex moral philosophical concept that is intimately tied to ideas about what it means to be human and serves as a basis for various obligations and rights. It assumes that human beings have certain capacities that allow them to make self-willed decisions and that they should be respected for this capacity (see chapter 7).

Machine autonomy, on the other hand, generally refers to the ability of machines to operate for extended periods of time without human intervention. They are delegated tasks to perform without a human operator continuously adjusting the behavior of the system. Such tasks can include navigating an aircraft, driving on the highway, buying stock, or monitoring a manufacturing process. Most of the time, this kind of machine autonomy concerns certain well-defined processes that can be fully automated. Autonomy is then the same as the high-end of a sliding scale of automation and has nothing to do with free will or similar moral philosophical concepts.

Admittedly, the distinction between machine autonomy and human autonomy starts to blur in discussions about AI technologies that learn and adapt to their environments. Systems that automate well-defined processes are relatively predictable, whereas systems that learn from their experiences and exhibit behaviors going beyond their original programming can seem like they have a "mind of their own." AlphaZero, a system designed to play games like chess and Go, is a frequently touted example of a system able to teach itself to play like a pro without explicit human guidance and in a way that is incomprehensible to its human developers. Can human beings still be meaningfully responsible for the behavior of such systems?

Well, despite the impressive feats of a system like this, its actions still take place within the constraints set by its human operators. It takes considerable expertise and effort from human developers to make the system operate properly. They have to carefully construct, adjust, and fine-tune the algorithms and select and prepare its training data.[16] The point is that various human agents are still in a position to exert some kind of control over the system, and with that comes some degree of responsibility. And, of course, it goes without saying that AlphaZero doesn't know that it's playing Go, or that it's being watched, or what a "game" is, or what it means to "play," or what it means to win, and it doesn't do anything other than play the games it was designed to play.

This isn't to say that the development of increasingly powerful AI won't pose real challenges to established ways of allocating responsibility. The increasing complexity of AI technologies and the growing distance between the decisions of developers and the outcomes of their choices will challenge existing ideas about who should be responsible for what and to what extent. And that's very much the point. The issue isn't whether human responsibility will cease to make any sense at all; it's about *which* human beings should be responsible and *what* their responsibility will look like. With the introduction of a new technology comes a redistribution of tasks, which will cause a shift of responsibility between the different actors in the chain. There may also be many more hands involved than before, as human-machine systems become more complex and larger. Some of these actors may become less powerful and will have less discretionary space, whereas others will gain more leverage. In any case, human beings are still making decisions about how to train and test the algorithms, when to employ them, and how to embed them in existing practices. They determine what is acceptable behavior and what should happen if the system transgresses the boundaries of acceptable behavior.

Figuring out who should be responsible for what and how isn't easy, and there are definitely bad ways of doing it. The opposite worry to the responsibility gap is that the *wrong* people will be held responsible. They become what Madeleine Elish calls a "moral crumple zone." These human actors are deemed responsible even though they have very limited or no control over an autonomous system. Elish argues that, given the complexity of these systems, the media and broader public tend to blame the nearest human operator such as pilots or maintenance staff rather than the technology itself or decision makers higher up in the decision chain.[17]

The advent of AI and the challenges it poses to existing ways of allocating responsibility raise a number of important questions. How can we properly distribute responsibility around increasingly complex technologies? What does it mean to be in control of an autonomous system? If the driver of an autonomous vehicle is no longer in charge of what the car does, on whom (or *what*) should responsibility fall? The distribution of responsibility isn't, after all, something that just happens, like a tree that falls in the woods in a way determined by the forces of nature (gravity, humidity, wind, etc.). It's something that we have to actively negotiate among ourselves, scoring off competing claims, always with an eye to the wider social ramifications of any settlement reached.

Negotiating Responsibilities

If the human driver of an autonomous vehicle is no longer actively controlling its behavior, can they be reasonably held liable, or should liability shift to the manufacturer or perhaps some other actor? If the vehicle's software learns from the environment and other road users,* could the manufacturer still be held liable for accidents caused by the vehicle? Are the existing rules for liability still applicable or should they be amended? Manufacturers' liability has often been strict. And, interestingly, strict liability was first recognized for the actions of substances (like dangerous chemicals) or chattels (like wandering sheep) that caused harm without the knowledge of their controller—a situation that might be thought directly parallel to the "mindless" but independent actions of machine learning algorithms that learn to classify and execute procedures without being explicitly programmed to do so by their developers. Is strict liability for manufacturers the way to go then?

One answer to the questions raised by the prospect of fully self-driving cars is that we won't need to come up with new laws, as the existing ones—including fault-based and strict liability regimes—will be sufficient to cover them. Manufacturers, it's supposed, are best placed to take precautions against the risks of harm in the construction phase and to warn users of those risks at the point of sale (or wholesale). Product liability rules should therefore apply as usual. No ifs, no buts.

Several legal scholars, however, argue that existing product liability schemes are insufficient and place an undue burden on the victims of accidents. Remember, even with strict product liability, it is the victim who must prove both that the product was defective and that the defect caused their harm. With the growing complexity of computer-enabled vehicles, it will become increasingly difficult for victims to show that the product was defective.[18] Product liability is already used sparingly because it's often difficult to show where an error originated. It may be easier to hold manufacturers liable in cases where it's clear that a particular problem occurs more often in particular designs, as when it came to light that one of Toyota's models would sometimes suddenly accelerate.[19] Even though engineers couldn't pinpoint the source of the malfunction, Toyota accepted liability given the high number of incidents. But in areas where there isn't such a high incidence rate, it can be very difficult to establish the liability of the manufacturer, strict or otherwise.

*Not that any autonomous car so far developed *can* learn on its own. To date, all autonomous vehicles ship from the factory with an algorithm learned in advance. That algorithm will probably have used training data gathered from real cars on real roads, but learning will have happened under the supervision of a team of engineers.

Moreover, like other complex systems, autonomous cars exacerbate the problem of many hands. *Multiple* actors will be involved in the manufacturing and functionality of the vehicle. Although the manufacturer of the actual physical artifact may be primarily in control of the physical artifact, other actors will contribute to the operation of the vehicle, such as various software developers, the owner of the car who needs to keep the software up to date, and the maintenance agency in charge of sensors on the road. This ecosystem of technologies, companies, government departments, and human actors makes it difficult to trace where and why mishaps occurred, especially when they involve technologies that self-learn from their environments.

In light of these difficulties, other legal scholars have championed various forms of compensation schemes that don't require proof of fault or even a human actor. Maurice Schellekens argues that the question of "who is liable for accidents involving self-driving vehicles" might be redundant. He points out that several countries, including Israel, New Zealand, and Sweden, already have no-fault compensation schemes (NFCS) in place for automobile accidents. In these countries, the owner of a car takes out mandatory insurance (or is covered by a more general state scheme that includes personal injury arising from road incidents), and when an accident occurs, the insurer will compensate the victim, even if no one is at fault or even *caused* the accident. Imagine cruising on the highway between Sydney and Canberra and swerving suddenly to avoid a kangaroo that jumped straight in front of your car—not an altogether uncommon occurrence along that stretch of road. Most likely there'd be no one to blame for any ensuing crash (much as in our previous example of the bee sting). An NFCS would at least enable the victim to be compensated quickly for any injuries sustained without the rigamarole of court proceedings.

A similar approach may be appropriate, Schellekens suggests, for accidents involving autonomous vehicles. Whether manufacturers will be incentivized to design safe vehicles under such a regime will depend on the rights of victims or insurers to pursue manufacturers for any defects that may have caused these incidents. For instance, a victim may be compensated by their insurer, but if faulty design was the issue, the insurer should acquire the victim's rights to sue the manufacturer in their place, or the victim themselves may pursue the manufacturer for compensation above the statutory (NFCS) limit. Another option would be to force *manufacturers* to pay the cost of insurance, rather than private citizens. Either way, it's important that manufacturers be held responsible for the harms caused by their designs, otherwise they'll have little incentive to improve them.

Morally and Legally Responsible AI?

The challenges that AI poses to existing ways of distributing responsibility have led some to suggest that we should rethink who or what we consider to be the responsible agent. Should this always be a human agent, or can we extend the concept of agency to include nonhuman agents? As we noted earlier, conventional moral philosophical notions of responsibility are roundly individualistic and anthropocentric. Only humans can be morally responsible. Legal conceptions of responsibility, on the other hand, allow for more flexibility here, as the agent doesn't necessarily have to be an individual human being, nor do they have to be in direct control of events.

In view of the advent of AI, some philosophers have argued that the human-centered conception of moral responsibility is outdated.[20] The complexity of certain kinds of software and hardware demands a different approach, one where artificial agents can be dealt with directly when they "behave badly." Other philosophers have also argued that if these technologies become complex and intelligent enough, they could well be attributed moral agency.[21]

Critics of this suggestion have countered that such an approach diminishes the responsibility of the people that develop and deploy autonomous systems.[22] They are human-made artifacts and their design and use reflect the goals and ambitions of their designers and users. It's through human agency that computer technology is designed, developed, tested, installed, initiated, and provided with instructions to perform specified tasks. Without this human input, computers could do nothing. Attributing moral agency to computers diverts our attention away from the very forces that shape technology to behave as it does.

One response to this would be to argue that, although technology on its *own* might not have moral agency, moral agency itself is also hardly ever "purely" human anyway.[23] According to Peter-Paul Verbeek, for example, human action is a composite of different forms of agency at work: the agency of the human performing the action; the agency of the designer who helped shape the mediating role of the artifact; and the artifact itself mediating between human actions and their consequences.[24] Whenever technology is used, moral agency is rarely concentrated in a single person; it is diffused in a complex mash-up of humans and artifacts.

In legal scholarship, a similar issue has been discussed in terms of personhood. Given the complexity of AI and digital ecosystems, some legal scholars as well as policy makers have suggested that artificial agents should be attributed some kind of personhood and be held liable the way corporations may be. In 2017, the European Parliament even invited the European Commission to consider the creation of a specific legal status that would establish electronic personhood for sufficiently sophisticated and autonomous robots and AI systems.

In law, however, the idea of nonhuman personhood is hardly exotic, as many legal systems already recognize different kinds of personhood.[25] Legal personhood is a legal fiction used to confer rights and impose obligations on certain entities like corporations, animals, and even rivers like the Ganges and Yamuna in India, the Whanganui in New Zealand, or whole ecosystems in Ecuador. Needless to say, the reasons for granting such status to these kinds of entities are entirely different from those justifying our view of each other as persons. Human beings are assigned rights and obligations based on moral considerations about, for example, their dignity, intrinsic worth, consciousness, autonomy, and ability to suffer. Where nonhuman entities are afforded personhood, it is for a great range of reasons, such as sharing certain human characteristics like the ability to suffer in the case of animals, or for economic reasons in the case of corporations. Many legal systems recognize corporations as persons to reduce the risk associated with individual liability and thereby encourage innovation and investment. AI technologies could perhaps be considered a kind of corporation. Indeed, an NFCS, extended (if necessary) to cover harms arising from the use of autonomous systems, is arguably already halfway there.

Looking Ahead

We've seen how being morally responsible for your actions goes along with *having* control over them, in some sense. Much the same is true when we use technology to achieve our ends. Being responsible for the outcomes of our actions when they have been facilitated by technology requires that we have some degree of control over that technology. This is because when we work with technology, the technology becomes a part of us—an extension of our arms and legs and minds. Although we've said a little about control in this chapter, our next chapter will elaborate on some of the ways in which technology inherently *resists* being controlled by us. This is important because the more control an individual has over a system, the more we can hold them responsible for the results of its deployment. On the other hand, the less meaningful and effective that control, the weaker our authority to blame them when things go wrong.

5 Control

Just about everybody who champions artificial intelligence agrees that there's a limit to how far AI should be allowed to usurp human jurisdiction.[1] That limit, understandably enough, is human destiny. Humanity should be able decide for itself what the ends of human life are. This is so, even though one of the most obvious things about ends is that humans rarely seem to agree on what they should be (beneath the level of platitudes, of course). The idea here is that AI should never *ultimately* replace human wishes, in some deep sense of "ultimate." So, even if people drawn from different demographics and geographically widely-dispersed cultures don't see eye-to-eye on the big questions, most of these same people would agree that it should in some sense be *up to people* to devise answers to these questions.

This chapter isn't about values or theories of value, human flourishing, and the "life well lived." But it does proceed on the reasonable assumption that, in an important sense, humans should always ultimately be in charge of what happens to them, that even if we decide to use labor-saving, creativity-liberating, and error-minimizing technology, we'll do so on *our* terms—that is, only so long as the technology conforms to our wishes. After all, that's what it means for a system to be under ultimate human control—for it to behave the way it should, the way we *want* it to behave, even if we aren't in moment-by-moment "operational" control of the system (and let's face it, why *would* we have operational control of an "autonomous" system?). The corollary of this setup, of course, is that if the system goes rogue, it's our prerogative to "switch it off."

This desire for humans to be in charge gets cashed out in various ways, but probably the most popular catchphrase at the moment—especially in discussions about the military deployment of lethal autonomous weapons—is the call for "meaningful human control." This is a more demanding requirement than ultimate control. Having ultimate control doesn't mean that we can prevent mishaps or calamities in the event something goes horribly wrong with an AI. It doesn't even necessarily mean that we can mitigate the worst results in a crisis. All it really means is that we have the power to "switch off" a system that has gone rogue to prevent further damage. But by this stage, the worst might have already happened. By contrast, for an autonomous system to be under *meaningful* control, something stronger than ultimate control is required (otherwise what would make it "meaningful"?). We think meaningful control implies that the system is under *effective* control, so that operators *are* in a position to prevent the worst from happening and thereby mitigate or contain the potential fallout. So, whereas ultimate control allows us merely to reassert our hegemony over an AI to which we've voluntarily ceded operational control, *meaningful* control means that we can reassert this hegemony *effectively*, that is, in sufficient time to avert a catastrophe (for example). We agree that this is a standard worth aspiring to, and from here on in we're going to assume that humans shouldn't give up meaningful control over an autonomous system, and certainly never when the stakes are high.

This simple rule of thumb, however, isn't always easy to adhere to in practice, and the main obstacle is probably psychological. In this chapter, we're going to take a look at some of the problems identified by researchers in the field of industrial psychology known as "human factors." What human factors research reveals is that in some situations—essentially, when autonomous systems reach a certain threshold of reliability and trustworthiness—to relinquish operational control *is* to relinquish meaningful control. Put simply, once humans are accustomed to trust a system that's reliable *most* of the time (but not *all* of the time), humans *themselves* tend to "switch off," falling into a sort of "autopilot" mode where diffidence, complacency and overtrust set in. Humans with this mindset become far less critical of a system's outputs, and, as a result, far less able to spot system failures. Telling ourselves that we retain "ultimate" control in these circumstances because we can always just press the "stop" button if we want to is a little bit delusional (it must be said).

The challenge is vividly illustrated by lethal autonomous weapons systems (LAWS). In the LAWS literature, one finds a tentative distinction drawn between humans being "*in* the loop," "*on* the loop," and "*off* the loop." Being "*in* the loop" means a human gets to call the shots (literally) and say whether a target should be tracked and engaged. The buck stops squarely with the human who decides to attack. A system with a human "*on* the loop," like a drone, would identify and track a target but not actually engage without human approval. On the other hand, if the drone handled *everything*, from identifying and tracking all the way through to engaging a target (without human intervention), the system would be fully autonomous. In that event, the human authority would be "*off* the loop."

These alternatives represent a sliding scale of possibilities. The human factors issues we mentioned arise somewhere between humans being *on* and *off* the loop—that is, a point where human agents are still technically *on* the loop but so disengaged and inattentive that they might as well be considered *off* the loop. In the LAWS context, it's not hard to imagine what the loss of meaningful control could entail. An autonomous weapon that is ill-advisedly trusted to discriminate between enemy combatants and civilians or to apply only such force as is necessary in the circumstances (the principle of "proportionality" in the law of armed conflict) could cause unspeakable devastation. Think how easy it would be for an autonomous weapon to misclassify a child rushing toward a soldier with a stick as a hostile combatant. Remember the wolves and huskies problem from the prologue? Translated to the sphere of war, object classifier errors are no laughing matter. And without meaningful human control over such systems, who's to say these horrifying possibilities wouldn't materialize?

We'll leave discussion of LAWS behind now, though, because they raise a host of other issues too niche to be addressed in a general chapter on control (e.g., what if the possibility of "switching off" an autonomous weapon—the one protocol we should *always* be able to count on in an emergency—also fails us because an enemy has hacked our console? Should the decision on who gets to live or die ever be left to a machine? Etc.). Instead, let's consider some more humdrum possibilities of autonomous systems getting out of control. Take criminal justice, a forum in which machine learning systems have been taken up with real panache, assisting in everything from police patrol and investigations to prosecution decisions, bail, sentencing, and parole. As a recent French report into artificial intelligence notes, "it is far easier for a judge to follow the recommendations of an algorithm that presents a prisoner as a danger to society than to look at the details of the prisoner's record himself and ultimately decide to free him. It is easier for a police officer to follow a patrol route dictated by an algorithm than to object to it."[2] And as the AI Now Institute remarks in a recent report of its own, "[w]hen [a] risk assessment [system] produces a high-risk score, that score changes the sentencing outcome and can remove probation from the menu of sentencing options the judge is willing to consider."[3] The Institute's report also offers a sobering glimpse into just how long such systems can go without being properly vetted. A system in Washington, D.C., first deployed in 2004 was in use for fourteen years before it was successfully challenged in court proceedings. The authors of the report attributed this to the "long-held assumption that the system had been rigorously validated."[4] In her book, *Automating Inequality*, Virginia Eubanks notes the complacency that high tech decision tools can induce in the social services sector. Pennsylvania's Allegheny County introduced child welfare protection software as part of its child abuse prevention strategy. The technology is supposed to assist caseworkers in deciding whether to follow up on calls placed with the County's child welfare hotline. In fact, however, Eubanks relates how caseworkers would be tempted to adjust their estimates of risk to align with the model's.[5]

When complaints have been made about these systems—and here we mean *formal* complaints, indeed in some of the highest courts of appeal—the remarks of judges often suggest that the full scale of the challenge hasn't really sunk in. Of all the uses of algorithmic tools in criminal justice, perhaps the most scrutinized and debated has been COMPAS (which we introduced in chapter 3). First developed in 1998 by the company Northpointe (now Equivant), COMPAS is used by criminal justice agencies across the USA.[6] In 2016, Eric Loomis launched an unsuccessful legal challenge to the use of COMPAS in the determination of his sentence. Without going into the details here, what's interesting is that, of all the concerns the appeal court expressed about the use of such tools, it obviously *didn't* think the control issue was a major one—judging by how casually the court's provisos on the future use of COMPAS were framed.

The court noted that sentencing judges "must explain the factors in addition to a COMPAS risk assessment that independently support the sentence imposed. A COMPAS risk assessment is only one of many factors that may be considered and weighed at sentencing."[7] The court also required that sentencing judges be given a list of warnings about COMPAS as a condition of relying on its predictions.[8] These warnings draw attention to controversies surrounding use of the tool and the original motivations behind its development—it was primarily devised as an aid to post-sentencing decision-making (like parole) rather than sentencing per se.

But that's it! The scale of the human factors challenge apparently hadn't dawned on anyone. (And why would it? They're judges, not psychologists!) A sentencing judge might well fish around for, and believe themselves to be influenced by, other factors "that independently support the sentence imposed." But if the control problem is taken seriously—in particular, the problem of "automation complacency" and "automation bias," which we'll describe in a moment—this strategy offers only false reassurance. The warnings themselves were mild. And even if they were more pointed, the truth is that we simply don't know if this sort of guidance is enough to knock a judge out of their complacency. Some research suggests that even explicit briefings about the risks associated with the use of a particular tool won't mitigate the strength of automation bias.[9]

Our worries aren't academic. Warning a judge to be skeptical of an automated system's recommendations doesn't tell them *how* to discount those recommendations. It's all very well being told that a system's recommendations are not foolproof and must be taken with a pinch of salt, but if you don't know *how* the system functions and *where* and *why* it goes astray, what exactly are you supposed to do? At what point *and in what way* is a judge meant to give effect to their skepticism? Research in cognitive psychology and behavioral economics (some of it discussed in chapters 2 and 3) also points to the effects of "anchoring" in decision-making. Even weak evidence can exert an unwholesome influence on a decision maker who is trying to be objective and fair. A judge that is given a high-risk score generated by a machine with fancy credentials and technical specifications may lean toward a higher sentence unwittingly under the force of anchoring. We just don't know if warnings and a duty to take other factors into account are powerful enough, *even in combination*, to negate such anchoring effects.

The Control Problem Up Close

What we're calling "the control problem" arises from the tendency of the human agent within a human-machine control loop to become complacent, over-reliant, or overtrusting when faced with the outputs of a reliable autonomous system. Now at first blush this doesn't seem like an insurmountable problem, but it's harder to manage than it seems.

The problem was first recognized in the 1970s,[10] but it didn't receive a definitive formulation until a little paper came along in 1983 with the succinctly telling title: "Ironies of Automation." The author was Lisanne Bainbridge, and the chief irony she grappled with was this, "that the more advanced a control system is, so the more crucial may be the contribution of the human operator."[11] Although writing at a time before deep learning had anything to do with algorithmically automated decision tasks, what she had to say about the role of the human agent in a human-machine system still rings true today.

> If the decisions can be fully specified, then a computer can make them more quickly, taking into account more dimensions and using more accurately specified criteria than a human operator can. There is therefore no way in which the human operator can check in real-time that the computer is following its rules correctly. *One can therefore only expect the operator to monitor the computer's decisions at some meta-level, to decide whether the computer's decisions are "acceptable."*[12]

As we see things, this residual monitoring function of the human operator generates at least four kinds of difficulties that should be treated separately (see box 5.1). The first relates to the cognitive limits of human processing power (the "capacity problem"). As Bainbridge put it, "If the computer is being used to make the decisions because human judgment and intuitive reasoning are not adequate in this context, then which of the decisions is to be accepted? The human monitor has been given an impossible task."[13]

Humans are often at a severe cognitive disadvantage vis-à-vis the systems they are tasked with supervising. This can be seen very clearly in the case of high-frequency financial trading. It's impossible for a monitor to keep abreast of what's happening in real time because the trades occur at speeds that vastly exceed the abilities of human monitors to keep track. As Gordon Baxter and colleagues point out, "[i]n the time it takes to diagnose and repair [a] failure ... many more trades may have been executed, and possibly have exploited that failure."[14] Similar issues arise from the use of autopilot systems in aviation that are becoming "so sophisticated that they only fail in complex 'edge cases' that are impossible for the designers to foresee."[15]

The second difficulty relates to the *attentional* limits of human performance (the "attentional problem").

> We know from many "vigilance" studies ... that it is impossible for even a highly motivated human being to maintain effective visual attention toward a source of information on which very little happens, for more than about half an hour. This means that it is humanly impossible to carry out the basic function of monitoring for unlikely abnormalities. ...[16]

Box 5.1
Breaking Down the Control Problem

The control problem breaks down into four more basic problems:

1. The Capacity Problem

Humans aren't able to keep track of the systems they're tasked with supervising because the systems are too advanced and operate at incredible speeds.

2. The Attentional Problem

Humans get very bored very quickly if all they have to do is monitor a display of largely static information.

3. The Currency Problem

Use it or lose it. Skills that aren't regularly maintained will decay over time.

4. The Attitudinal Problem

Humans have a tendency to overtrust systems that perform reliably *most* of the time (even if they are not reliable *all* of the time).

Automation has a significant impact on situation awareness.[17] For example, we know that drivers of autonomous vehicles are less able to anticipate take-over requests and are often ill prepared to resume control in an emergency.[18]

The third difficulty relates to the *currency* of human skills (the "currency problem"). Here is Bainbridge, again: "Unfortunately, physical skills deteriorate when they are not used. ... This means that a formerly experienced operator who has been monitoring an automated process may now be an inexperienced one."[19]

The fourth and final difficulty, and the one we've chosen to focus on in this chapter, relates to the *attitudes* of human operators in the face of sophisticated technology (the "attitudinal problem"). Except for a few brief remarks,[20] this problem wasn't really addressed in Bainbridge's paper.[21] It has, however, been the subject of active research in the years since.[22] Here the problem is that as the quality of automation improves and the human operator's role becomes progressively less demanding, the operator "starts to assume that the system is infallible, and so will no longer actively monitor what is happening, meaning they have become complacent."[23] Automation complacency often co-occurs with automation *bias*, when human operators "trust the automated system so much that they ignore other sources of information, including their own senses."[24] Both complacency and bias stem from *overtrust* in automation.[25]

What makes each of these problems especially intriguing is that each gets worse as automation *improves*. The better a system gets, the more adept at handling complex information and at ever greater speeds, the more difficult it will be for a human supervisor to maintain an adequate level of engagement with the technology to ensure safe resumption of manual control should the system fail. When it comes to the current ("SAE Level 2") fleet of autonomous vehicles* that allow the driver to be hands- and feet-free (but not *mind*-free, because the driver still has to watch the road), legendary automotive human factors expert Neville Stanton expressed the conundrum wryly: "Even the most observant human driver's attention will begin to wane; it will be akin to watching paint dry."[26] And as far as complacency and bias go, there is evidence that operator trust is directly related to the scale and complexity of an autonomous system. For instance, in low-level partially automated systems, such as SAE Level 1 autonomous vehicles, there is "a clear partition in task allocation between the driver and vehicle subsystems."[27] But as the level of automation increases, this allocation gets blurred to the point that drivers find it difficult to form accurate assessments of the vehicle's capabilities, and on the whole are inclined to overestimate them.[28]

These results hold in the opposite direction too. *Decreases* in automation reliability generally seem to *increase* the detection rate of system failures.[29] Starkly put, automation is "most dangerous when it behaves in a consistent and reliable manner for most of the time."[30] Carried all the way, then, it seems the only safe bet is to use dud systems that don't inspire overtrust, or, on the contrary, to use systems that are provably better-than-human at particular tasks. The latter option is feasible because once a system is provably better (meaning less error prone) than a human agent at performing a particular task, an attentive human supervisor over that system will be superfluous. Whether there's a human keeping watch over the system or

*The Society of Automotive Engineers (SAE) framework, running from Level 0 (no automation) to Level 5 (full automation) classifies vehicles in accordance with the degree of system functions that have been carved out for automation (SAE J3016 2016). Tesla Autopilot and Mercedes Distronic Plus (Level 2) require the driver to monitor what is going on throughout the whole journey, whereas Google's self-driving car does everything except turn itself on and off.

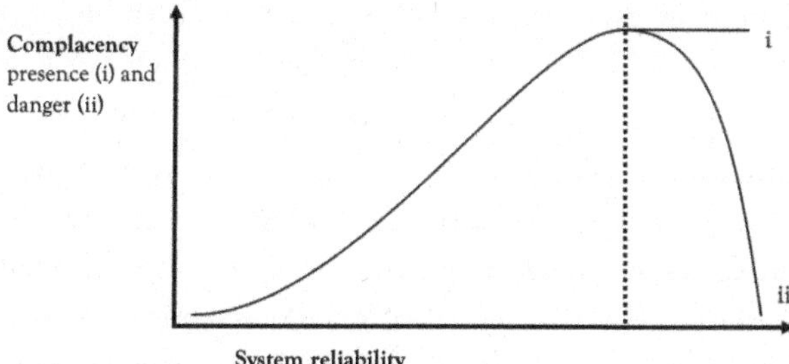

Figure 5.1
Presence (i) and danger of complacency (ii) as a function of system reliability. The dashed line represents near perfect (better-than-human) reliability.

not—and if there is, whether the human succumbs to automation complacency or not—it won't really matter, because the system will have fewer error rates than the poor human struggling to keep up.

Figure 5.1 depicts both the presence and danger of complacency as a function of system reliability. Notice that at a certain point of reliability (represented by the dashed line), the presence of complacency no longer matters because it will no longer pose a danger.

Using "Better-Than-Human" Systems: Dynamic Complementarity

If the question is interpreted literally, the answer to "Can the control problem be solved?" appears to be straightforwardly negative. The control problem can't *literally* be solved. There is nothing we can do, as far as we know, that *directly* targets, much less directly curbs, the human tendency to fall into automation complacency and bias once an autonomous system operates reliably most of the time, and when the only role left for the operator is to monitor its largely seamless transactions. However, by accepting this tendency as an obstinate feature of human-machine systems, we might be able to work around it without pretending we can alter constraints imposed by millions of years of evolution.

The insights of human factors research are instructive here. One important human factors recommendation is to foster mutual accommodation between human and computer competencies through a *dynamic* and *complementary* allocation of functions. Humans should stick to what they do best, such as communication, abstract reasoning, conceptualization, empathy, and intuition; computers can do the rest.[31] At the same time, the allocation should be flexible enough to support *dynamic* interaction whenever it would contribute to optimal performance (or is otherwise necessary), with hand-over and hand-back for some tasks, for instance, when a driver disengages cruise control and thereby resumes control of acceleration. This assumes that some decisions can be safely handled by both humans and computers and that humans and computers have shared competencies within particular subdomains. Hand-over and hand-back may also go some way toward alleviating the currency problem, as operators are thereby afforded an opportunity to practice and maintain their manual control skills.

The obvious assumption behind this approach is that decision tasks can be cut more or less finely. We can assume, for instance, that *border control* (i.e., whether to admit, or not admit, persons moving between state boundaries) is one big decision involving customs clearance, passport verification, drug detection, and so on. We can also assume that either: (a) the *entire* border control decision is handled by one large, distributed border control software package, or (b) that only some automatable subcomponents of the overall decision have been carved out for discrete automation, the rest being left to human controllers. Currently, of course, border control decisions are only partially automated. SmartGate allows for fully automated electronic passport control checks, but customs officials still litter most immigration checkpoints. And that's the point—their job is to handle only those parts of the overall decision that can't be efficiently automated. Maybe one day the whole decision chain *will* be automated, but we're not there yet.

Under a regime of dynamic and complementary allocation, obviously some autonomous systems will replace human agents and be left to operate without supervision. Human-machine systems that contain automated subcomponents work best when the human operator is allowed to concentrate their energies on chunks of the task better suited to human rather than autonomous execution. But obviously—as we've already intimated—this setup only avoids the control problem if the automated subroutines are handled by systems approaching near-perfect (better-than-human) dependability. Otherwise the autonomous parts might work very well most of the time but still require a human monitor to track for occasional failures—and it's clear where *this* path leads.

It's reasonable to ask, though: how many autonomous systems actually reach this threshold? Truth is, it's difficult to say. SAE Level 2 (and higher) autonomous vehicles certainly don't yet approach this kind of reliability.[32] But many subcomponents within standard (nonautonomous, SAE Level 0) vehicles clearly do, such as automatic transmission, automatic light control, and first-generation cruise control.[33]

In more typical decision support settings, arguably diagnostic and case prediction software are approaching this better-than-human standard. For instance, there are AI systems that can distinguish between lung cancers and give prognoses more accurately than human pathologists armed with the same information, and systems that can spot Alzheimer's with 80 percent accuracy up to a decade before the first appearance of symptoms, a feat surely outperforming the best human pathologists.[34] In the legal sphere, advances in natural language processing and machine learning have facilitated the development of case prediction software that can predict, with an average 79 percent accuracy, the outcomes of cases before the European Court of Human Rights when fed the facts of the cases alone.[35] Most impressively, a similar system had better luck predicting the rulings of the US Supreme Court than a group of eighty-three legal experts, almost half of whom had previously served as the justices' law clerks (60 percent vs. 75 percent accuracy).[36] Beyond these reasonably clear-cut cases we can only speculate. One advantage of dynamic complementarity is precisely that, by carving up a big decision into smaller and smaller chunks, the more likely we'll be able to find a better-than-human system to take up the baton.

And what if the sort of better-than-human accuracy we have in mind here can't be assured? The upshot of our discussion is that a decision tool shouldn't replace a human agent, at least in a high-stakes/safety-critical setting, unless the tool reaches a certain crucial threshold of reliability. But what if this standard can't be met? Can less-than-reliable systems be deployed? The short answer is yes. As we've noted, the control problem doesn't arise from the use of patently suboptimal automation, only from *generally* dependable automation. So, depending on the circumstances, a *less*-than-reliable system might safely replace a human agent charged with deciding some matter within a larger decision structure (e.g., passport verification within the larger border control decision structure). The problem with a tool like COMPAS (among other problems) is that it straddles the line between reliability in particular settings and overall optimality. It's not reliable enough to meet the better-than-human standard, but it's still useful in some ways. In other words, it's exactly the kind of tool liable to induce automation complacency and bias, and it does so in a high-stakes setting (bail, sentencing, and parole decisions).

Are There Other Ways to Address the Control Problem?

There is some evidence that increasing accountability mechanisms can have a positive effect on human operators whose primary responsibility is monitoring an autonomous system. An important study found that "making participants accountable for either their overall performance or their decision accuracy led to lower rates of automation bias."[37] This seems to imply that if the threat of random checks and audits were held over monitors, the tendency to distrust one's own senses might be attenuated. What effects these checks could have on other aspects of human performance and job satisfaction is a separate question, as is the question of how accountability mechanisms affect automation *complacency* (as opposed to *bias*). More creative accountability measures, such as "catch-trials," in which system errors are deliberately generated to keep human invigilators on their toes, could also be useful in counteracting automation bias. Catch-trials are quite popular in aviation training courses. The aviation industry is actually a fine example of how to manage automation bias, since automation bias is a life-threatening problem in this arena, and taken very seriously. But in any case, much like other touted solutions to the control problem, they don't offer a literal, *direct*, solution. Rather, they render systems that are mostly dependable (but not better than human) *less reliable by stealth* (as it were), capitalizing on the premise that less reliable systems don't induce the same complacency and bias that attend more reliable systems.

What about teamwork? Might having a *group* of humans in the loop, working together and keeping watch on one another, alleviate automation bias? Apparently not.

> Sharing monitoring and decision-making tasks with an automated aid may lead to the same psychological effects that occur when humans share tasks with other humans, whereby "social loafing" can occur—reflected in the tendency of humans to reduce their own effort when working redundantly within a group than when they work individually on a given task. … Similar effects occur when two operators share the responsibility for a monitoring task with automation.[38]

Finally, guidelines recommending that decision makers exercise their own judgment *before* consulting an algorithm could assist in offsetting some of the effects of automation complacency and bias. In these cases, the algorithm would serve merely as a check on a decision maker's intuitions. Note that this approach comes close to telling decision makers *not* to use the algorithm. So again, it's not so much a *solution* to the problem—a way of directly targeting and curbing an ingrained psychological bent—as it is a way of managing, negotiating, and (in this case) *avoiding* the problem.

The Take-Home Message

Automation introduces more than just automated parts; it can transform the nature of the interaction between human and machine in profound ways. One of its most alarming effects is to induce a sense of complacency in its human controllers. So among the factors that should be considered in the decision to automate any part of an administrative or business decision is the tendency of human operators to hand over meaningful control to an algorithm just because it works well in most instances. It's this problem, not machines taking over per se, that we really have to watch out for.

6 Privacy

Ask people what they think of when they hear the phrase "breach of privacy," and you'll get some predictable responses. Common among these might be examples of police or military surveillance, closed circuit television cameras watching your every move, paparazzi snooping into private family moments with long camera lenses, or a government agency intercepting personal messaging services. These examples fit with the understanding most people have of their right to privacy. Most people would probably take the right to mean something like "the right to be let alone."[1] But there's more to it than that, and in this chapter we want to help you better understand the different dimensions of privacy, how new technologies affect these dimensions, why this matters, and how these issues relate to you as a citizen.

Dimensions of Privacy

It might come as a surprise that, given the cross-cultural importance of privacy, there is still no internationally settled definition of what it actually is. In 1948, members of the United Nations adopted the Universal Declaration of Human Rights, which set out the fundamental human rights they agreed must be universally protected, including the right to privacy in Article 12: "No one shall be subjected to arbitrary interference with his privacy, family, home or correspondence, nor to attacks upon his honor and reputation. Everyone has the right to the protection of the law against such interference or attacks."[2]

Unfortunately, as fine a sentiment as this is, countries have been free to interpret this standard as they've seen fit. Europe has perhaps traveled farthest along the path to securing robust privacy rights for its citizens. Most jurisdictions haven't come nearly as far. In any case, trying to secure international consensus on the meaning of the term isn't the only problem. Getting *anyone* to define the concept (let alone over 150 countries) is a tall order.

The privacy scholar Julie Inness once lamented that privacy is a concept in chaos. One of the entries in the index to her book on privacy reads, "Quagmire, privacy as deep."[3] But chaotic though it may be, Inness still felt it was salvageable—a concept whose many jagged fragments could be made to fit together.[4] This is because, even if privacy has a broad range of meanings, not all of which are mutually accommodating, all of them do seem to share a preoccupation with intimacy and dignity. For Inness, privacy boils down to our natural concern for maintaining a degree of autonomy with respect to matters we care about, such as intimacy in our relationships.

Years later, the legal scholar Daniel Solove would try his own hand at a definition, but unlike Inness, he chafed at essentialism. Taking his cue from the famous Austrian philosopher Ludwig Wittgenstein, Solove doubts we'll ever find the unique, irreducible essence of a concept like privacy. As far as he's concerned, we're better off abandoning this search altogether, redirecting our attention to the concrete consequences of privacy violations. Accordingly, he devised a taxonomy of privacy *harms*. In his scheme there are harms that arise from information collection (such as surveillance), harms that arise from information processing (such as aggregation, identification, or insecure processing), harms arising from information dissemination (such as unlawful disclosure or breach of confidence), and harms arising from invasion (such as physical intrusion).[5] A harms-based approach is certainly useful, focusing on what's likely to matter most to people when they think about their privacy. But Solove's particular approach also raises questions the moment we bring machine learning into the picture. If datasets don't relate to anyone in particular—they are usually anonymized aggregations of many individual profiles run together—and if they aren't likely to be used against any of the individuals whose data go into them, then what exactly is the harm? A predictive model can, of course, be weaponized against someone if the algorithm is stacked against them, for example, if by using a particular software package a person's gender or ethnicity makes it almost certain they'll be rejected for a job. But it isn't obvious that the harm here results from a breach of privacy per se. It certainly doesn't seem to be a breach of *that person's* privacy (at least not straightforwardly). And even if your personal information *is* being funneled back "against" you, what exactly is the "harm" if all you're having to contend with are Netflix and Amazon recommendations?

This isn't to say that focusing on harms is misguided, but it is a reminder to make sure any working definition takes our brave new world of big data seriously.

Here, we're going to be pragmatic. We're not going to pretend privacy means just one thing or results in one basic kind of harm (and nor did Solove, for that matter). Instead, there are at least four things privacy can mean—four *dimensions* of privacy, if you will—each of which results in a distinctive kind of harm:

1. **Bodily privacy** secures a person's *bodily integrity* against nonconsensual touching or similar interference;
2. **Territorial privacy** protects a person's *ambient space* from intrusion and surveillance;
3. **Communication privacy** protects a person's *means of communicating* against interception;
4. **Informational privacy** prevents *personal information* being collected, processed, or used against its owner's wishes (otherwise known as "data protection").[6]

This "divide and conquer" strategy makes clear how new forms of privacy breach can emerge in a particular dimension. For example, modern informational privacy (data protection) principles arose partly in response to the vast and rapidly expanding amounts of personal information that both governments and private corporations held about individuals. Concerns over the adequacy of existing national laws and international standards resulted in litigation and, ultimately, new laws imposing clear obligations on agencies holding personal information.* Such agencies must now be careful to ensure that individuals who give them personal information retain power over that information, including the power to access, correct, use, and (if necessary) delete it. New forms of communication privacy have emerged too. Immediately following the Snowden revelations, for example, legal scholars and civil libertarians debated whether there could be a right to privacy with respect to electronic communications.

*For example, in 1980 the German Constitutional Court ruled on the validity of population census data, holding that in the age of data processing, individual rights to self-determination required protection from the unlimited collection, use, and storage of personal information.

Although each of these privacy dimensions is important—and AI has implications for every one of them—in this chapter, we aren't going to discuss bodily, territorial, or communication privacy in much detail. Instead, our focus will be on informational privacy, as this is the form of privacy most obviously and directly threatened by the advent of big data. Besides, advances in AI and machine learning may mean that these four dimensions will converge increasingly in years to come. It's a safe bet that ever larger and more sophisticated datasets will be used to enhance both state and corporate powers of surveillance and intrusion. To take just one example, face recognition software is already a kind of surveillance technology, and it depends crucially on access to high quality training data.

Informational Privacy and AI

The overarching concern about AI and privacy arises from the ways in which predictive algorithms and other forms of machine learning can recognize patterns in our personal information, perhaps even ones that we ourselves cannot see. In doing so, these technologies can peer into parts of our private lives that no other human would be able to see. This is, again, the *overarching* concern. There are more specific concerns too, of course, the two most important of which we'll consider here. But it's worth keeping the big picture in view as we look at these more specific problems.

A fundamental principle of data protection law is that "data should be collected for named and specific purposes."[7] But purpose limitation strikes at the heart of the big data business model. As we've seen, developing the capabilities of machine learning techniques requires access to training data—lots and lots of training data (see chapter 1). Privacy advocates worry in particular about *how* personal information is collected, especially on the internet. It is rarely used for "named and specific purposes," and rarely given with informed consent.[8] Oh, sure, companies like Facebook *do* mention purposes, but they are often so general as to be meaningless; for example, to provide and improve their products and services—in other words, to implement Facebook!

Knowing whether users who visit websites to book flights will next visit websites to book accommodation and then other websites to book rental vehicles is valuable information. Hotels and rental car companies are willing to pay for this information because it can be used to push targeted advertisements to travelers. Individuals are often peacefully unaware that the knowledge of what they do on the internet is valuable and that they are allowing others to collect and profit from the use of this information. And even when they know or suspect that their information will be sold to third parties, they won't know *exactly* how those third parties will use it, and indeed *who* those third parties are. The cynical interpretation is that people are being exploited. "If it's free on the internet, *you're* the product," as the saying goes.

The internet isn't the only data collection point of interest. Data can be gathered from what might seem to be the most unlikely places. Many washing machines are now part of the "internet of things," and come equipped with sensors that can generate information about wash times, wash cycles, and other matters that can be downloaded and used to predict maintenance, repairs, faults, and energy consumption.[9] Although such data collection may be beneficial from a maintenance and design point of view—knowing that most people use only two or three of the wash functions on a machine might lead to a more efficient design, for example—it might also be valuable to third parties. The local energy company (say) might be interested to encourage you not to use your washing machine during peak electricity usage hours. A manufacturer of laundry detergent might be interested in hawking related products, such as fabric softeners. Again, these third parties will be willing to pay for this information, and, no less than before, consumers will be frequently unaware of the kinds of data being collected about them, how much data are being collected, and what uses they're being put to.

It's true that many companies are open about the terms of the deal they have struck with users: "If you allow us to mine your likes, shares and posts—and to sell what we learn to the highest bidder—we'll grant you access to our platform."* But not all instances of data mining occur with explicit consent. And even in cases where they do, this "consent" isn't necessarily free and voluntary. If you need to "consent" to data collection to access an essential

---

*Interestingly, Facebook doesn't directly sell any data—they're not "data brokers." Instead, they monetize their data by offering a *targeted ad placement* service to advertisers. They engage in analytics to home in on the most suitable customer groups for specific products, and then allow advertisers to place ads with these specific groups. It's as if Facebook were to say to a travel agency, "I know all the middle-aged Christian train enthusiasts. I won't tell you who they are, but if you pay me, I'll pass your message on to them."

service, like an online banking or health app, in what sense are you meaningfully consenting to the surrender of your personal information? You don't have a choice—you *have* to consent in order to access the essential service. Besides, long, complex terms and conditions that no one has the time to read and that don't clearly set out "named and specific purposes" make a mockery of genuinely free and informed consent.

The second consent-related issue thrown up by big data concerns the use of *inferred* data. Inferred or "derived" data (such as the suburb you live in derived from your postcode) can be distinguished from *collected* data (data that you explicitly, knowingly, provide, such as your name) and *observed* data (data that you passively, or implicitly, provide, such as your handwriting, accent, or keystroke rate). Machine learning and big data literally *exist* to facilitate the drawing of inferences. When a machine learning tool flags a submitted tax return as potentially fraudulent, for example, it's not doing so through directly ascertainable information (fraud per se), but on the basis of significant correlations it has learned between directly ascertainable information (reported income, losses, etc.) and the phenomenon of interest (i.e., fraud). In a reliable dataset of previous tax return information, larger-than-usual losses reported in consecutive tax years—itself directly ascertainable information—might be found to correlate strongly with known instances of tax fraud. In the language of chapter 1, tax fraud would be the "predicted variable," the inferred characteristic (like mortality given age, or weight given height). The question for privacy law is whether consent must be obtained for the use of this inferred information. Do inferred data count in the same way as primary (collected and observed) data?

This isn't just an academic worry. Consider machine learning techniques that use nonsensitive information to predict very sensitive information about an individual. Privacy advocates have raised concerns about the use of apparently unrelated data, such as information about location, social media preferences, screen time on different apps or phone activity, to aggregate and predict highly sensitive information (such as sexuality or political beliefs). One study found that the emotional states of computer users could be determined from such apparently innocuous information as their keystroke rate.[10] In 2017, a Stanford University study claimed that an algorithm could successfully distinguish between gay and straight men 81 percent of the time, and between gay and straight women 71 percent of the time. The study used deep neural networks to extract features from over thirty-five thousand photographs—facial features that many of us would probably regard as nonsensitive (our faces are, after all, the one part of us always open for the world to see). The authors concluded that "given that companies and governments are increasingly using computer vision algorithms to detect people's intimate traits, our findings expose a threat to the privacy and safety of gay men and women."[11] There has been considerable skepticism about these particular findings on methodological grounds,[12] but follow-up models correcting various aspects of the original still appear to have some ability to identify sexual orientation from photographs.[13]

Beyond these two paramount data protection issues that machine learning has brought to the fore, a range of other, somewhat less definitive issues have also arisen in recent years. One is the potential for reidentification from anonymized datasets. Anonymization is used frequently in academia to protect the privacy of experimental subjects. It's easy to forget that psychological and medical experiments don't just happen. Among a myriad of other precautions and ethical protocols, researchers have to recruit willing participants, a task made marginally easier by being able to guarantee subjects that their personal information won't end up in the wrong hands. But scholars like Paul Ohm have emphasized just how weak anonymization techniques have become. Using publicly available information, Ohm says that, despite efforts to anonymize them, it's possible to make highly accurate predictions about the identity of specific individuals in a dataset.[14] This does not augur well for scientific research. Reidentification can be very easy. For example, you might suffer from a rare condition or live in a rural area where the number of people fitting your characteristics is low. In other cases in which reidentification may be more difficult because your personal information is not so uniquely identifiable, reidentification is still surprisingly technically straightforward. One French study found that 75 percent of mobile phone users within a dataset could be reidentified based on an individual's use of just two smart phone apps. The reidentification rate increased to 90 perecent if four, rather than two, smart phone apps were used.[15]

There are also questions about how existing data protection standards might apply to personal information used for machine learning and big data analytics. These questions relate to how individuals can access their information, who controls the information for the purposes of liability for correction, exactly *where* obligations for ensuring accuracy fall in the data chain (particularly when information is repurposed or passed to a third party), what obligations for data deletion should be imposed, and how long data can be retained before it should be deleted. Only some of these questions have received answers in some jurisdictions. And even then, the answers aren't always decisive.

AI, Privacy, and the Consumer

So how, in practice, will all this affect your privacy and the privacy of your family, friends, and others in your community? What should your expectations of privacy be? And given how little knowledge or power you might feel you have, will your expectations of privacy matter anyway?

First up, let's consider the targeted ads we mentioned earlier. These ads are focused on almost every aspect of your online life, from that weekend away you just had to those new shoes you'd like to buy. They are a key part of online sales and product and service promotion, and used by a vast array of companies worldwide. Why does this particular application of machine learning create consumer privacy risks? There are consent-related objections to it, as we saw. But is that all?

One way in which targeted ads pose consumer privacy risks is in their potential for discrimination. The fact is that these practices aren't solely aimed at affecting your selection of a product or service from a range of choices. The techniques used to promote these can also affect *whether you are offered certain choices in the first place.*

Take housing. In early 2019, the US government filed a lawsuit against Facebook, alleging that its targeted advertising algorithms violated the Fair Housing Act by discriminating against some people, using data mining practices to restrict which users were able to view its housing-related ads.[16] The Fair Housing Act makes discrimination in housing and housing-related services illegal. For example, it's illegal to advertise that housing is available to people of a certain race, color, national origin, religion, sex, or marital status. Ben Carson, Secretary of the Department for Housing and Urban Development, put it simply: "Facebook is discriminating based on who people are and where they live. Using a computer to limit a person's housing choice can be just as discriminatory as slamming a door in someone's face."[17] This effect can be amplified if those offering targeted ads have significant market dominance. Facebook is estimated to control about 20 percent of online advertising in the United States.*

Public lawsuits are helpful because they enable us to monitor at least some of what is happening and to have judicial oversight of practices that affect consumers. But commentators point out that recent lawsuits against Facebook by the National Fair Housing Alliance, American Civil Liberties Union, and other civil society groups have been settled out of court, and, in some of those cases, the terms of the settlements remain private. This makes it harder to figure out the specific privacy-protective measures, if any, that such litigation is forcing companies to take.[18]

Now you might think that this whole business of online discriminatory advertising isn't itself *squarely* a privacy issue. It's really a cocktail of anti-discrimination, fair trading, and human rights issues jumbled together. Still, let's not forget that discriminatory advertising is *discriminating* (in the sense of *discerning*), based on technology that predicts the kind of person you *are*—technology, indeed, that infers things about you that you may not want others to know. In some countries, being gay is illegal and punishable by death. "Gaydar" software that "outs" you at an airport security checkpoint wouldn't merely be inconvenient; it could be life-threatening.

Let's consider another way in which machine learning and natural language processing tools in particular are using information in ways that affect your privacy as a consumer: training data that are used to develop consumer behavior prediction tools. This is clearly a privacy issue. Our behaviors, intentions, and innermost proclivities are being predicted, and possibly even manipulated (see chapter 7). As we've discussed, those developing machine learning tools rely on existing datasets to train and test them. These datasets range in size, quality, and diversity, and are used in many different ways depending on the type of AI being developed. So where does the data come from? You might be surprised to learn that you have probably already given your information to a training dataset. For example, if you've ever made a call to your insurance, phone, or electricity company,

---

*Similar action is taking place in other countries. In 2018, for instance, Privacy International filed complaints with UK, French, and Irish data protection authorities against seven companies, complaining of their use of personal data for targeted advertising and exploitative purposes.

you've probably contributed to the collection of this kind of training data. You might recall an irritating automatically-generated voice message, telling you that your call "may be recorded for quality and training purposes." Imagine hundreds and thousands, perhaps millions, of those calls being made available to train natural language processing tools that recognize all kinds of useful things: differences between male and female voices, when someone is angry, when they are upset, the typical questions customers ask, and the typical complaints they make.

Thinking of canceling your insurance policy and switching to another provider? Based on your behavior, machine learning tools can anticipate this. Customer "churn prediction" or attrition rates are prominent performance indicators in many companies, and being able to predict and reduce likely churn can provide a significant business advantage. The models can be trained on the behavioral data of thousands of previous customers who have switched accounts and those who have stayed. Using this information to predict likely customer churn, a list of "at risk" customers can be generated and sent to an accounts manager for review and action.

That might be well and good—perhaps you've been missing out on a better deal and really appreciate that call from the insurance company to see if you're still happy with their product. But when these kinds of datasets are used to create profiles of different types of people based on preselected categories of information, such as age, sex, medical history, location, family status, and so on, mispredictions will be rife. What if the profile generated for you is so wildly different from your actual situation that you miss out on some options completely?* What if, like Virginia Eubanks, your family's health insurance account is red-flagged by an algorithm for suspicious behavior because when your domestic partner was assaulted you (very reasonably) made a health insurance claim for domestic care services around the same time you switched jobs and took out a new policy? We know that algorithms can infer lots of uncanny things about you. This creepiness factor is one thing when the inferences are correct but quite another when the inferences are wrong.[19]

---

*You can check your Facebook profile to see what categories you've been placed in. Go to Settings > Privacy Shortcuts > More Settings > Ads > Your Information > Review and Manage Your Categories. You'll see some true stuff, and probably some weird stuff.

Companies are also combining dynamic price differentiation with data collection to enable algorithmic setting of real time service prices. For example, a report by Salesforce and Deloitte in 2017 found that although uptake of algorithms by major brand businesses was still low—with just over a third of such businesses adopting AI—among those that have taken up algorithmic tools, 40 percent were using them to tailor prices.[20]

Why is this important? Well, you might think you're roaming freely online, privately seeking out the information you want when and how you want it. In fact, your online life is increasingly being curated, filtered, and narrowed. In the process, not only is your sphere of privacy reducing, but the breadth of your participation in public life is also reducing, simply because your freedom to receive information of *any* kind is being restricted (see chapter 7). This doesn't just affect the opportunities and choices available to you online, either. It's increasingly affecting your offline existence, too, including at work (see chapter 9). One privacy implication is the detrimental effect of being constantly monitored while working. This is another domain where, as we mentioned earlier, different dimensions of privacy (here the territorial and informational dimensions) are converging. Your personal information can be used to help employers infer the movements and habits of "workers like you." The difference between being "watched" by a camera and being "known" by an algorithm is becoming less important.

AI, Privacy, and the Voter

In the first part of 2017, a news story broke about the UK's referendum on leaving the EU—a story that would change the landscape of political campaigning in the United Kingdom and around the world. Media reports emerged in the United Kingdom that Cambridge Analytica and related companies (which we'll just collectively call "Cambridge Analytica" for ease of reference) had assisted the Leave campaign by providing data services that supported micro-targeting of voters. Eighteen months later, in November 2018, the UK Information Commissioner, Elizabeth Denham, reported on her investigation of the story during which she had engaged 40 investigators who identified 172 organizations and 71 witnesses, issued 31 notices demanding information, executed two warrants, instigated one criminal prosecution and seized materials including 85 pieces of equipment, 22 documents and 700 terabytes of data (the equivalent of more than 52 billion pages of evidence).[21] The investigation uncovered how political campaigns use individuals' personal information to target potential voters with political messages and advertisements. It revealed the complex system of data sharing between data brokers, political parties and campaigns, and social media platforms. The commissioner concluded: "We may never know whether individuals were unknowingly influenced to vote a certain way in either the UK EU referendum or in the US election campaigns. But we do know that their personal privacy rights have been compromised by a number of players and that the digital electoral eco-system needs reform."[22]

Again a key concern was the use of data collected for one purpose for a completely different purpose without consent and in violation of Facebook policies. The commissioner found that Cambridge Analytica worked with university researcher and developer, Dr. Aleksandr Kogan, to establish a company (GSR) that contracted with Cambridge Analytica to develop a new app, *thisisyourdigitallife*. Users who logged into Facebook and authorized the app made their data—*and* those of their Facebook friends—available to GSR and Cambridge Analytica. (To prevent this sharing, the friends would have had to uncheck a field in their Facebook profile that was on by default, something hardly any users ever did.) The new app was able to access about 320,000 Facebook users who took detailed personality tests while logged into their Facebook accounts. In doing so, the app was able to collect the users' public profile (including birth date, current city, photos in which the user was tagged, pages they had liked, timeline and newsfeed posts, lists of friends, email addresses, and Facebook messages). Facebook estimated that the total number of users of the app, including affected Facebook friends, was approximately 87 million.[23]

Should We Give Up on Privacy?

With all of these concerns it is fair to ask, is privacy dead? Contrary to popular belief, demand for privacy will likely increase rather than decrease in the age of AI. Many surveys show that consumers value their online privacy. In the United States in 2015, Consumer Reports found that 88 percent of people regard it as important that no one is listening to or watching them. Echoing this finding, a Pew research study found that the majority of Americans believe it is important, or very important, to be able to maintain their privacy in everyday life. In relation to online life, 93 percent of adults said that being in control of who can get information about them is important, and 95 percent believed that what is collected about them is important. However, the same survey found that nearly two thirds of people were not confident that their online activities would be kept secure by online advertisers, social media sites, search engine providers, or online video sites.[24]

Global civil society organizations such as Privacy International and the Electronic Frontier Foundation, as well as consumer rights groups, have long called on users to take more control of the collection and use of their personal information online. They have also advocated for simpler privacy protections. The fact is there *are* ways to limit the effects of online advertising, but surprisingly few implement them. Sometimes it's because companies make it hard. If you don't consent to your phone call being recorded, you might not be able to access a service. In other cases, even if you do take care with your online activities—for example, by staying away from certain platforms or not posting information about your political beliefs, health status, or social activities—algorithms can still predict what you might do next. Before the "friend's permission" feature in Facebook was disabled, your personal information may have been stored in your friends' social media accounts and therefore vulnerable if your friends chose to allow other apps to access their contacts. Another significant barrier is the lack of simple tools to help consumers.

Can We Protect Privacy in the Age of AI?

There are now much blurrier lines between personal and nonpersonal information, new forms of data exploitation, new forms of data collection, and new ways in which personal information is being used to create profiles and predictions about individuals through machine learning tools. The result is that the various dimensions of privacy we've discussed are both expanding and constricting in exciting, complex, and confusing ways. In light of all of this, you might well ask, can we better protect individual privacy? Taking a little more responsibility for our privacy settings is one step, but what else can be done?

Calls have been made to strengthen privacy in technology development. Ann Couvakian coined the phrase "privacy by design" to help conceptualize the request to technology developers for privacy-enhancing technologies.[25] The EU's General Data Protection Regulation (GDPR) charted new territory in limiting the use of automated processing in certain circumstances and requiring individuals to be provided with information as to its existence, the logic involved and the significance and proposed consequences of the processing for the individual concerned. The GDPR also confers rights of correction and erasure.

Remember that machine learning's predictive accuracy depends on the future looking like the past (see chapter 1). If the data on which an algorithm has been trained and tested remain the same, the predictions will also remain the same, and that's fine. But data are rarely static. People change. They develop new skills, end relationships, form new ones, change jobs, find new interests. And some people will always fall through the cracks of what is considered typical for someone of a certain age, gender, sexuality, ethnicity, and so on. The result is that inferences can be based on data that are obsolete or in some other way "dirty." But given how pervasive and persistent inferential analytics is set to become, a key step would be to develop the law here even further—beyond the modest protections offered by the GDPR. What can we allow to be reasonably inferred, in what sorts of situations and under what kinds of controls (such as rights of access, correction, and challenge, which the GDPR already confers to some degree)? Recently, a right to reasonable inferences has been proposed to reorient data protection law toward the *outputs* of personal information as distinct from its collection and use (the traditional focus of data protection law).[26] The proposal would require data controllers to show why particular data are relevant to drawing certain "high risk" inferences, why the inferences themselves need to be drawn at all, and whether both the data and methods used to draw these inferences are statistically reliable.

Legal systems also need to come clean on the status of inferred data—should it attract the same protections as the primary (collected and observed) personal information on which it is based? It's been suggested that inferred data *can* be defined as personal information if the content, purpose, or result of the data processing relates to an identifiable individual.[27]

We've touched on issues of freedom and voluntariness in this chapter. In our next chapter, we'll probe more deeply into the ways that both legal and illegal uses of your personal information potentially compromise your freedom as a human agent.

7 Autonomy

Autonomy is an important value in liberal societies. It is protected and cherished in both legal and popular culture. Some people argue that a life without it would not be worth living. In his speech to the Second Virginia Convention on March 23, 1775, Patrick Henry is said to have persuaded people to support the American Revolutionary War with a rousing speech that ended with the line, "Give me liberty or give me death." Similar thoughts are echoed in the New Hampshire state motto with its bold imperative to "live free or die." And it's not just in the United States that these sentiments find a welcome audience. The Greek national motto is "Eleftheria i Thanatos," which translates as "liberty or death." Furthermore, commitment to individual liberty and autonomy has often been a source of solace for those who find themselves in difficult circumstances. Nelson Mandela, for example, consoled himself and other inmates during his imprisonment on Robben Island by reciting William Ernest Henley's poem "Invictus." Written in the late 1800s while Henley was recovering from multiple surgeries, the poem is a paean to self-mastery, independence, and resilience, closing with the immortal lines, "I am the master of my soul / I am the captain of my fate."

Given the cherished status of individual autonomy in a liberal society, we must ask how AI and algorithmic decision-making might impact upon it. When we do, we find that there is no shortage of concern about the potentially negative impacts. Social critics of the technology worry that we will soon become imprisoned within the "invisible barbed wire" of predictive algorithms that nudge, manipulate, and coerce our choices.[1] Others such as the Israeli historian Yuval Noah Harari argue that the ultimate endpoint for the widespread deployment of AI is a technological infrastructure that replaces and obviates rather than nudges and manipulates autonomous human decision makers.[2] But are they right? Is autonomy really imperiled by AI? Or could AI be just the latest in a long line of autonomy-enhancing technologies?

In this chapter, we step back from the hype and fear-mongering and offer a more nuanced guide to thinking about the relationship between AI and individual autonomy. We do this in four stages. First, we clarify the nature and value of autonomy itself, explaining what it is and how it is reflected in our legal systems. Second, we consider the potential impact of AI on autonomy, asking in particular whether the technology poses some novel and unanticipated threat to individual autonomy. Third, we ask whether the negative impacts of AI on autonomy are more likely to emanate from the private sector or the public sector, or some combination of the two. In other words, we ask, "Who should we fear more: big tech or big government?" And fourth, we consider ways in which we can protect individual autonomy in the era of advanced AI.

The overall position put forward in this chapter is that, although we must be vigilant against the threats that AI poses to our autonomy, it is important not to overstate those threats or assume that we, as citizens, are powerless to prevent them from materializing.

A Three-Dimensional Understanding of Autonomy

In order to think properly about the impact of AI on autonomy, we need to be clear about what we understand by the term "autonomy." The opening paragraph mingled together the ideals of autonomy, liberty, independence, and self-mastery. But are these all the same thing or are there important differences between them? Philosophers and political theorists have spent thousands of years debating this question. Some of them have come up with complex genealogies, taxonomies, and multi-dimensional models of what is meant by terms like "liberty" and "autonomy."[3] It would take several books to sort them all out and figure out who provides us with the best understanding of the relevant terms. Rather than do that, what we propose to do in this chapter is offer one specific model of autonomy. This model is inspired by long-standing philosophical debates, and those familiar with those debates will be able to trace the obvious influences, but you won't require any prior knowledge of them to follow the discussion. The goal is to provide readers with an understanding of autonomy that is reasonably straightforward and self-contained but sufficiently nuanced to enable them to appreciate the multiple different impacts of AI on autonomy.

So what is this model of autonomy? It starts by acknowledging the common, everyday meaning of the term, which is that to be "autonomous" one must be able to choose one's own path in life, engage in self-rule, and be free from the interference and manipulation of others. This common understanding is a good starting point. It captures the idea that to be autonomous requires some basic reasoning skills and abilities—specifically, the ability to pick and choose among possible courses of action—and some relative independence in the exercise of those skills and abilities.

The legal and political philosopher Joseph Raz adds some further flesh to the bones of this model of autonomy in a famous definition of what it takes to be autonomous.

> If a person is to be maker or author of his own life, then he must have the mental abilities to form intentions of a sufficiently complex kind, and plan their execution. These include minimum rationality, the ability to comprehend the means required to realize his goals, the mental faculties necessary to plan actions, etc. For a person to enjoy an autonomous life he must actually use these faculties to choose what life to have. There must in other words be adequate options available for him to choose from. Finally, his choice must be free from coercion and manipulation by others, he must be independent.[4]

Raz's definition breaks autonomy down into three component parts. It says that you are autonomous if (1) you have the basic rationality required to act in a goal-directed way; (2) you have an adequate range of options to choose from; and (3) your choice among those options is independent, that is, free from coercion and manipulation by others. This definition can be applied to both individual decisions and lives as a whole. In other words, following Raz, we can consider whether a person's whole life is autonomous or whether a particular decision or set of decisions is autonomous. Some people like to use different terms to differentiate between the different scopes of analysis. For instance, the philosopher Gerald Dworkin once suggested that we use the term "autonomy" when talking about life as a whole (or some extended portion of life) and "freedom" when talking about individual decisions. This, however, is confusing because, as we shall see below, the term "freedom" has also been applied just to Raz's third condition of autonomy. So, in what follows, we'll just use the one word—"autonomy"—to refer to the phenomenon in which we are interested. Furthermore, we won't really be discussing the impact of AI on life as a whole but rather its impact on specific decisions or specific decision-making contexts.

It's tempting to think about Raz's three components of autonomy as conditions that need to be satisfied in order for a decision to count as autonomous. If you satisfy all three, then you are autonomous; if you do not, then you are not. But this is too binary. It's actually more helpful if we think about the three components as *dimensions* along which the autonomy of a decision can vary. We can call these the rationality, optionality, and independence dimensions, respectively. There may be a minimum threshold that needs to be crossed along each dimension before a decision will qualify for autonomous status, but beyond that threshold decisions can be more or less autonomous.

To make this more concrete, compare two different decisions. The first decision involves you choosing among movie options that are recommended to you by your Netflix app (or other video streaming app). The app gives you ten recommended movies. You read the descriptions of them and choose the one in which you are most interested. The second decision involves you choosing a walking route that has been recommended to you by Google Maps (or some other mapping service). The app gives one recommended route, highlighted in blue, and one other similar route in a less noticeable grey. You follow the highlighted route. Are both of these choices autonomous? Probably. We can assume that you have the basic rationality needed to act in a goal-directed way; we can see that you have been given a range of options by the apps; and the apps do not obviously coerce and manipulate your choices (though we'll reassess this claim below). But is one decision more autonomous than the other? Probably. On the face of it, it seems like the Netflix app, by giving more options and information and not highlighting or recommending one in particular scores higher along the second and third dimensions. It gives you more optionality and more independence.

Thinking about autonomy in this three-dimensional way is useful. It allows us to appreciate that autonomy is a complex phenomenon. It encourages us to avoid simplistic and binary thinking. We can now appreciate that autonomy is not some simple either/or phenomenon. Decisions can be more or less autonomous, and their autonomy can be undermined or promoted in different ways. In this respect, it is important to realize that there may be practical tradeoffs between the different dimensions and that we need to balance them in order to promote autonomy. For example, adding more options to a decision may only promote autonomy *up to a point.*

Beyond that point, adding more options might be overwhelming, make us more confused, and compromise our basic rationality. This is something that psychologists have identified as "the paradox of choice" (more on this later).[5] Similarly, some constraints on options and some minimal coercion or interference might be necessary to make the most of autonomy. This has long been recognized in political theory. Thomas Hobbes, for example, in his famous defense of the sovereign state, argued that we need some minimal background coercion by the state to prevent us from sinking into bitter strife and conflict with one another.

Although all aspects of the three-dimensional model of autonomy deserve scrutiny, it is worth focusing a little more attention on the independence dimension. This dimension effectively coincides with the standard political concepts of "freedom" and "liberty." The key idea underlying this dimension is that, in order to be autonomous, you must be free and independent. But what does that freedom and independence actually entail? Raz talks about it requiring the absence of coercion and manipulation, but these concepts are themselves highly contested and can manifest in different ways. Coercion, for instance, usually requires some threatened interference if a particular option is not chosen ("do this or else!"), whereas manipulation can be more subtle, often involving some attempt to brainwash or condition someone to favor a particular option. In modern political theories of freedom, there is also an important distinction drawn between freedom understood as the absence of *actual* interference with decision-making and freedom understood as the absence of *actual* and *potential* interference. The former is associated with classical liberal theories, such as those put forward by John Locke and Thomas Hobbes; the latter is associated with republican theories of freedom, such as those favored by Machiavelli and, more recently, the Irish philosopher Philip Pettit (not to be confused with the Republican party in the US).[6]

We can explain the distinction with a simple example taken from Pettit. Imagine a slave, that is, someone who is legally owned and controlled by another human being. Now imagine that the slave owner just happens to be particularly benevolent and enlightened. They have the legally recognized power to force the slave to do as they will, but they don't exercise that power. As a result, the slave lives a relatively happy life, free from any actual interference. Is the slave free? Proponents of the republican theory of freedom would argue that the slave is not free. Even though they are not actively being interfered with, they are still living in a state of *domination*. The benevolent slave master might only allow the slave to act within certain arbitrarily defined parameters, or they might change their mind at any moment and step in and interfere with the slave's choices. That's the antithesis of freedom. The problem, according to the republicans, is that the classical liberal view cannot account for that. We need to consider both actual and potential interferences with decision-making if we are to properly protect individual liberty. A person cannot be free if they live under the potential domination of another.

For what it's worth, we think that this is a sensible position to adopt. The bottom line, then, is that we need to ensure that our three-dimensional model of autonomy includes both actual and potential interferences in its approach to the independence dimension.

That's enough about the nature of autonomy. What about its value? Why would some people rather die than do without it? We can't exactly answer that question here, but we can at least map out some of the ways it *could* be answered. We'll leave it up to the individual reader to determine how important autonomy is to them.

Broadly speaking, there are two ways to think about the value of autonomy. The first is to think of autonomy as something that is *intrinsically* valuable, that is, valuable in itself, irrespective of its consequences or effects. If you take that view, then you'll think of autonomy as one of the basic conditions of human flourishing. You'll probably believe that humans cannot really live good lives unless they have autonomy. You may even go so far as to think that a life without autonomy is not worth living. The second way to think about it is as something that is *instrumentally* valuable, that is, valuable because of its typical consequences or effects. If you take this view, you might highly prize autonomy but only because you think autonomy is conducive to good outcomes. This is a common view and features widely in political and legal justifications for autonomy. The idea is that people are better off if they are left to do their own thing. They can pick and choose the things that are most conducive to their well-being rather than having the state or some third party do this for them. The counterpoint to this, of course, is the *paternalistic* view, which holds that people don't always act in their best interests and sometimes need a helping hand. The battle between paternalism and autonomy is rife in modern political debates. At one extreme, there are those who decry the "nanny state" for always stepping in and thinking it knows best; at the other extreme, there are those that lament the irrationality of their fellow citizens and think we would all become obese addicts if left to decide for ourselves. Most people probably take up residence somewhere between these two extremes.

It is possible, of course, to value autonomy for both intrinsic and instrumental reasons, to think that it is valuable in and of itself and more likely to lead to better outcomes. Even more exotic views are available too. For example, one of the authors of the present book has defended the claim that autonomy is neither instrumentally nor intrinsically valuable but rather something that makes good things better and bad things worse.[7] Imagine, for example, a serial killer who autonomously chooses to kill lots of people versus a serial killer who has been brainwashed into killing lots of people. Whose actions are worse? The obvious answer is the former. The latter's lack of autonomy seems mitigating. It is still bad, of course, for many people to have died as a result of his actions, but it seems worse, all things considered, if they died as a result of an autonomous choice. It suggests a greater evil is at work. The same is true for good things that happen accidentally versus those that happen as the result of autonomous action.

Whichever view you take, there are additional complexities to consider. For starters, you need to consider where in the hierarchy of values autonomy lies. We value lots of things in life, including health, sociality, friendship, knowledge, well-being, and so on. Is autonomy more or less important than these other things? Or is it co-equal? Should we be willing to sacrifice a little autonomy to live longer and happier lives (as the paternalists might argue)? Or do we agree with Patrick Henry that the loss of liberty is a fate worse than death? How we answer those questions will play a big role in how seriously we take any threat that AI may pose to our liberty.

An analogy might help. We discussed the importance of privacy in chapter 6. Lots of philosophers, lawyers, and political theorists think that privacy is important, but some people question our commitment to privacy. After all, the lesson of the internet age seems to be that people are quite willing to waive their right to privacy in order to access fast and efficient digital services. This leads some to argue that we might be transitioning to a post-privacy society. Could something similar be true for autonomy? Might we be willing to waive our autonomy in order to gain the benefits of AI? Could we be transitioning to a post-autonomy society? This is something to take seriously as we look at the potential threats to autonomy.

Finally, you should also consider the relationship between autonomy (however it is valued) and other basic legal rights and freedoms. Many classic negative legal rights, for example, have a firm foundation in the value of autonomy. Examples would include freedom of speech, freedom of movement, freedom of contract, freedom of association, and, of course, privacy and the right to be left alone. Although there are economic and political justifications for each these freedoms, they can also be justified by the value of autonomy, and their protection can help to promote autonomy. In short, it seems like a good case can be made for the view that autonomy, however it is valued, is foundational to our legal and political framework. We all have some interest in it.

Do AI and Algorithmic Decision Making Undermine Autonomy?

Now that we have a deeper understanding of both the nature and value of autonomy we turn to the question at hand: does the widespread deployment of AI and algorithmic decision-making undermine autonomy? As mentioned, critics and social commentators have already suggested as much, but armed with the three-dimensional model, we can undertake a more nuanced analysis. We can consider the impact of these new technologies along all three dimensions of autonomy.

Doing so, we must admit upfront that there is some cause for concern. Consider the first of the three dimensions outlined above: basic rationality. Earlier chapters in this book have highlighted various ways in which this might be affected by AI and algorithmic decision-making tools. Basic rationality depends on our ability to understand how our actions relate to our goals. This requires some capacity to work out the causal structure of the world and to pick actions that are most likely to realize our goals. One of the problems with AI is that it could prevent us from working these causal relationships out. Consider the problem of opacity and the lack of explainability we discussed in chapter 2. It's quite possible that AI systems will just issue us with recommended actions that we ought to follow without explaining why or how those options relate to our goals. We'll just have to take it on faith. In a recent book entitled *Re-Engineering Humanity,* Brett Frischmann and Evan Selinger articulated this fear in stark terms. They argue that one of the consequences of excessive reliance on AI would be the re-programming of humans into a simple "stimulus-response" machines.[8]

Humans will see the recommendations given to them by AI and implement the response without any critical reflection or thought. Basic rationality is compromised.

Similar problems arise when we look at the optionality dimension. One of the most pervasive uses of AI in consumer-facing services is to "filter" and constrain options. Google, for example, works by filtering and ordering possible links in response to our search queries. It thereby reduces the number of options that we have to process when finding the information we want. Netflix and Amazon product recommendations work in a similar way. They learn from your past behavior (and the behavior of other customers) and give you a limited set of options from a vast field of potential products. Often this option-filtering is welcome. It makes decisions more manageable. But if the AI system only gives us one or two recommendations, and gives one of them with a "95 percent" confidence rating, then we have to query whether it is autonomy preserving. If we have too few options, and if we are dissuaded from thinking critically or reflectively about them, then we arguably lose something necessary for being the authors of our own lives. Chapter 5 on control is really a particular illustration of this problem.

Finally, the independence dimension can also be compromised by the use of AI. Indeed, if anything, the independence dimension is likely to be the most obvious casualty of pervasive AI. Why so? Because AI offers many new opportunities for actual and potential interferences with our decision-making. Outright coercion by or through AI is one possibility. An AI assistant might threaten to switch off a service if we don't follow a recommendation, for example. Paternalistic governments and companies (e.g., insurance companies) might be tempted to use the technology in this way. But less overt interferences with decision-making are also possible. The use of AI-powered advertising and information management can create filter bubbles and echo chambers.[9] As a result, we might end up trapped inside technological "Skinner boxes," in which we are rewarded for toeing an ideological party-line and thereby manipulated into a set of preferences that is not properly our own. There is some suggestion that this is already happening, with many people expressing concern about the impact of AI-mediated fake news and polarizing political ads in political debates.[10] Some of these ads potentially exemplify the most subtle and insidious form of domination, what Jamie Susskind refers to as *perception-control*.[11] Perception-control is, literally, the attempt to influence the way we perceive the world. Filtering is the lynchpin of perception-control. The world is messy and bewildering, so our transactions with it have to be mediated to avoid our being swamped in detail. Either we do this sifting and sorting ourselves, or, more likely, we rely on others to do it for us, such as commercial news outlets, social media, search engines, and other ranking systems. In each case, someone (or *something*) is making choices about how much is relevant for us to know, how much context is necessary to make it intelligible, and how much we need to be told. Although in the past filtering was done by humans, increasingly it is being facilitated by sophisticated algorithms that appeal to us based on an in-depth understanding of our preferences. These preferences are reliably surmised through our retail history, Facebook likes and shares, Twitter posts, YouTube views, and the like. Facebook, for example, apparently filters the news it feeds you based on about 100,000 factors "including clicks, likes, shares [and] comments."[12] In the democratic sphere, this technology paves the way to active manipulation through targeted political advertising. "Dark" ads can be sent to the very people most likely to be susceptible to them without the benefit—or even the possibility—of open refutation and contest that the marketplace of ideas depends on for its functioning. The extent of perception-control that new digital platforms make possible is really a first in history, and likely to concentrate unprecedented power in the hands of a few big tech giants and law-and-order-obsessed state authorities (see below).[13]

On top of all this, the widespread use of mass surveillance by both government and big tech companies creates a new form of domination in our lives. We are observed and monitored at all times inside a digital panopticon, and although we might not be actively interfered with on an ongoing basis, there is always the possibility that we might be if we step out of line. Our position then becomes somewhat akin to that of a digital slave living under the arbitrary whim of our digital masters. This is the very antithesis of the republican conception of what it means to be autonomous.[14]

But there is another story to tell. It is easy to overstate the scare-mongering about AI and the loss of autonomy. We have to look at it from all sides. There are ways in which AI can promote and enhance autonomy. By managing and organizing information into useful packages, AI can help us sort through the complex mess of reality. This could promote rationality, not hinder it. Similarly, and as noted, the filtering and constraining of options, provided it is not taken too far, can actually promote autonomy by making decision-making problems more cognitively manageable. With too many options we get stuck and become unsure of what to do. Narrowing down the field of options frees us up again.[15] AI assistants could also protect us from outside forms of interference and manipulation, acting as sophisticated ideological spam filters, for example. Finally, and more generally, it is important to recognize that technologies such as AI have a tendency to increase our power over the world and enable us to do things that were not previously possible. Google gives us access to more useful (and useless!) information than any of our ancestors; Netflix gives us access to more entertainment; AI assistants like Alexa and Siri allow us to efficiently schedule and manage our time. The skillful use of AI could be a great boon to our autonomy.

There is also the danger of status quo bias or loss aversion to consider.[16] Often when a new technology comes onboard we are quick to spot its flaws and identify the threats it poses to cherished values such as autonomy. We are less quick to spot how those threats are already an inherent part of the status quo. We have become desensitized to those. This seems to be true of the threats that AI poses to our autonomy. It is impossible to eliminate all possible threats to autonomy. Humans are not perfectly self-creating masters of their own fates. We depend on our natural environment and on one another. What's more, we have a long and storied history of undermining one another's autonomy. We have been undermining rationality, constraining options, and ideologically manipulating one another for centuries. We have done this via religious texts and government diktats. But we have also created reasonably robust constitutional and legislative frameworks that protect against the worst of these threats. Is there any reason to think that there is something special or different about the threats that AI poses to autonomy?

Perhaps. Although threats to autonomy are nothing new, AI does create new modalities for realizing those threats. For example, one important concern about AI is how it might concentrate autonomy-undermining power within a few key actors (e.g., governments and big tech firms). Historically, autonomy-undermining power was more widely dispersed. Our neighbors, work colleagues, friends, families, states, and churches could all try to interfere with our decision-making and ideologically condition our behavior. Some of these actors were relatively powerless, and so the threat they posed could be ignored on a practical level. There was also always the hope that the different forces might cancel each other out or be relatively easy to ignore. That might no longer be true. The internet connects us all together and creates an environment in which a few key firms (e.g., Facebook,

Google, Amazon) and well-funded government agencies can play an outsized role in constraining our autonomy. What's more, on the corporate side, it is quite possible that all the key players have a strong incentive to ideologically condition us in the same direction. The social psychologist and social theorist Shoshana Zuboff, in her book, *The Age of Surveillance Capitalism*, argues that all dominant tech platforms have an incentive to promote the ideology of surveillance capitalism.[17] This ideology encourages the extraction of value from individual data and works by getting people to normalize and accept mass digital surveillance and the widespread use of predictive analytics. She argues that this can be done in subtle and hidden ways, as we embrace the conveniences of digital services and normalize the costs this imposes in terms of privacy and autonomy. The same is true when governments leverage AI. The Chinese government, for example, through its social credit system (which is facilitated through partnerships with private enterprises) uses digital surveillance and algorithmic scoring systems to enforce a particular model of what it means to be a good citizen.[18] The net result of the widespread diffusion of AI via the internet is that it gives a handful of actors huge power to undermine autonomy. This might be a genuinely unprecedented threat to autonomy.

AI could also erode autonomy in ways that are much more difficult for individuals to resist and counteract than the traditional threats to autonomy. The idea of the "nudge" has become widespread in policy circles in the past decades. It was first coined by Cass Sunstein and Richard Thaler in their book, *Nudge*.[19] Sunstein and Thaler were interested in using insights from behavioral and cognitive psychology to improve public policy. They knew that decades of research revealed that humans are systematically biased in their decision-making. They don't reason correctly about probability and risk; they are loss averse and short-termist in their thinking.[20] As a result, they often act contrary to their long-term well-being. Knowing all this seems like a justification for paternalistic intervention. Individuals cannot be trusted to do the right thing so someone else must do it for them. Governments, for example, need to step in and put people on the right track. The question posed by Sunstein and Thaler was "how can we do this without completely undermining autonomy?" Their answer was to nudge, rather than coerce, people into doing the right thing. In other words, to use the insights from behavioral psychology to gently push or incentivize people to do the right thing, but always leave them with the option of rejecting the nudge and exercising their own autonomy. This requires careful engineering of the "choice architectures" that people confront on a daily basis so that certain choices are more likely. Classic examples of "nudges" recommended by Sunstein and Thaler include changing opt-in policies to opt-out policies (thereby leveraging our natural laziness and loyalty to the status quo), changing how risks are presented to people to accentuate the losses rather than the gains (thereby leveraging our natural tendency toward loss aversion), and changing the way in which information is presented to people in order to make certain options more salient or attractive (thereby leveraging natural quirks and biases in how we see the world).

There has been much debate over the years about whether nudges are genuinely autonomy-preserving. Critics worry that nudges can be used manipulatively and nontransparently by some actors and that their ultimate effect is to erode our capacity to make choices for ourselves (since nudges work best when people are unaware of them). Sunstein maintains that nudges can preserve autonomy if certain guidelines are met.[21] Whatever the merits of these arguments, the regulatory theorist Karen Yeung has argued that AI tools facilitate a new and more extreme form of nudging, something she calls "hypernudging."[22] Her argument is that persistent digital surveillance combined with real time predictive analytics allows software engineers to create digital choice architectures that constantly adapt and respond to the user's preferences in order to nudge them in the right direction. The result is that as soon as the individual learns to reject the nudge, the AI system can update the choice architecture with a new nudge. There is thus much reduced capacity to resist nudges and exercise autonomy.

In a similar vein, AI enables a far more pervasive and subtle form of domination over our lives. Writing about this in a different context, the philosopher Tom O'Shea argues that there is such a thing as "micro-domination."[23] This arises when lots of small-scale, everyday decisions can only be made under the guidance and with the permission of another actor (a "dominus" or master). O'Shea gives the example of a disabled person living in an institutional setting to illustrate the idea. Every decision they make—including when to get up, when to go to the bathroom, when to eat, when to go outside, and so on—is subject to the approval of the caretakers employed by the institution. If the disabled person goes along with the caretakers' wishes (and the institutional schedule), then they are fine, but if they wish to deviate from that, they quickly find themselves unable to do what they want. Taken individually, these decisions are not particularly significant—they do not implicate important rights or life choices—but taken together all these "micro" instances of domination constitute a significant potential interference with individual autonomy.

The pervasive use of AI could result in something similar. Consider a hypothetical example: a day in the life of Jermaine. Just this morning Jermaine was awoken by his sleep monitoring system. He uses it every night to record his sleep patterns. Based on its observations, it sets an alarm that wakes him at the optimal time. When Jermaine reaches his work desk, he quickly checks his social media feeds where he is fed a stream of information that has been tailored to his preferences and interests. He is encouraged to post an update to the people who follow his work ("the one thousand people who follow you on Facebook haven't heard from you in a while"). As he settles into his work, his phone buzzes with a reminder from one of his health and fitness apps that it's time to go for a run. Later in the day, as he drives to a meeting across town, he uses Google Maps to plot his route. Sometimes, when he goes off track, it recalculates and sends him in a new direction. He dutifully follows its recommendations. Whenever possible, he uses the autopilot software on his car to save time and effort, but every now and then it prompts him to take control of the car because some obstacle appears that it's not programmed to deal with. We could multiply the examples, but you get the idea. Many small-scale, arguably trivial, choices in Jermaine's everyday life are now subject to an algorithmic master: what route to drive, whom to talk to, when to exercise, and so on. As long as he works within the preferences and options given to him by the AI, he's fine. But if he steps outside those preferences, he will quickly realize the extent of his dependence and find himself unable to do quite as he pleases (at least until he has had time to adjust to the new normal).

Indeed, this isn't a purely hypothetical concern; we already see it happening. The sociologist Janet Vertesi has documented how both she and her husband were flagged as potential criminals when they tried to conceal the fact that she was pregnant from online marketers who track purchasing data, keyword searches, and social media conversations.

> For months I had joked to my family that I was probably on a watch list for my excessive use of Tor and cash withdrawals. But then my husband headed to our local corner store to buy enough gift cards to afford a stroller listed on Amazon. There, a warning sign behind the cashier informed him that the store "reserves the right to limit the daily amount of prepaid card purchases and has an obligation to report excessive transactions to the authorities.[24]

She took one step outside the digital panopticon and was soon made aware of its power.

The net effect of both AI-mediated hypernudging and micro-domination is likely to be a form of learned helplessness. We might want to be free from the power of AI services and tools, but it is too difficult and too overwhelming to rid ourselves of their influence. Our traditional forms of resistance no longer work. It is easier just to comply.

All of this paints a dystopian picture and suggests that there might be something genuinely different about the threat that AI poses to autonomy. It may not pose a new category of threat, but it does increase the scope and scale of traditional threats.

But, again, some degree of skepticism is needed before we embrace this dystopian view. The threats just outlined are all inferred from reflections on the nature of AI and algorithmic power. They are not based on the careful empirical study of their actual effects. Unfortunately, there are not that many empirical studies of these effects just yet, but the handful that do exist suggest that some of the threats outlined above do not materialize in practice. For example, the ethnographer Angèle Christin has conducted in-depth studies of the impacts of both descriptive and predictive analytical tools in different working environments.[25] Specifically, she has looked at the impact of real-time analytics in web journalism and algorithmic risk predictions in criminal courts. Although she finds that there is considerable hype and concern about these technologies among workers, the practical impact of the technology is more limited. She finds that workers in both environments frequently ignore or overlook the data and recommendations provided to them by these algorithmic tools. Her findings in relation to risk prediction in criminal justice are particularly striking and important given the focus of this book. As noted in previous chapters, there is much concern about the potential bias and discrimination inherent in these tools. But Christin finds that very little attention is paid to them in practice. Most lawyers, judges, and parole officers either ignore them entirely or actively "game" them in order to reach their own preferred outcomes. These officials also express considerable skepticism about the value of these tools, questioning the methodology underlying them and usually favoring their own expert opinion over that provided to them by the technology.*

*In chapter 5, we saw that AI doesn't induce complacency when the technology is *perceived* to be suboptimal.

Similarly, some of the concerns that have been expressed about ideological conditioning, particularly in the political sphere, appear to be overstated. There is certainly evidence to suggest that fake news and misinformation is spread online by different groups and nation states.[26] They often do this through teams of bots rather than real human beings. But research conducted by the political scientists Andrew Guess, Brendan Nyhan, Benjamin Lyons and Jason Reifler suggests that people don't get trapped in the digital echo chambers that so many of us fear, that the majority of us still rely on traditional mainstream news media for our news, and in fact that we're more likely to fall into echo chambers *offline* than online![27]

Findings like these, combined with the possible autonomy-enhancing effects of AI, provide some justification for cautious optimism. Unless and until we are actively forced to use AI tools against our will, we have more power to resist their ill effects than we might realize. It is important that this is emphasized and that we don't get seduced by a narrative of powerlessness and pessimistic fatalism.

Big Tech or Big Government?

A minor puzzle emerges from the foregoing discussion. We have canvassed the various threats that AI poses to autonomy. We have made the case for nuance and three-dimensional thinking when it comes to assessing those threats. But we have been indiscriminate in how we have thought about the origin of those threats. Is it from big tech or big government that we have most to fear? And does it matter?

You might argue that it does. There's an old trope about staunch libertarians of the pro-business, anti-government type (the kind you often find in Silicon Valley): they are wary of all threats to individual liberty that come from government but seem wholly indifferent to those that come from private businesses.

On the face of it, this looks like a puzzling inconsistency. Surely all threats to autonomy should be taken equally seriously? But there is some logic to the double-standard. Governments typically have far more coercive power than businesses. While you can choose to use the services of a business, and there is often competition among businesses that gives you other options, you *have* to use government services: you can't voluntarily absent yourself from them (short of emigrating or going into exile). If we accept this reasoning, and if we review the arguments outlined above, we might need to reconsider how seriously we take the threats that AI poses to individual autonomy. Although some of the threats emanate from government, and we should be wary of them, the majority emanate from private enterprises and may provide less cause for concern.

But this is not a particularly plausible stance in the modern day and age. For one thing, there is often a cozy relationship between government and private enterprise with respect to the use of AI. This topic will be taken up in more detail in the next chapter, but for now we can simply note that governments often procure the services of private enterprises in order to carry out functions that impact on citizens' rights. The predictive analytics tools that are now being widely used in policing and criminal sentencing, for example, are ultimately owned and controlled by private enterprises that offer these services to public bodies at a cost. Similarly, the Chinese Social Credit System—which is probably the most invasive and pervasive attempt to use digital surveillance and algorithmic scoring to regulate citizens' behavior—is born of a cozy relationship between the government and the private sector. So it is not easy to disentangle big government from big tech in the fight to protect autonomy.

More controversially, it could be argued that at least some big tech firms have taken on such an outsized role in modern society that they need to be held to the same standards (at least in certain respects) as public bodies. After all, the double-standard argument only really works if there is reasonable competition among private services, and people do in fact have a choice of whether or not to use those services. We may question whether that is true when it comes to the goods and services offered by the tech giants like Google, Amazon, Facebook, and Apple. Even when there is some competition among these firms, they tend to sing from the same "surveillance capitalist" hymnbook.*

In a provocative paper entitled "If Data Is the New Oil, When Is the Extraction of Value from Data Unjust?"[28] the philosopher Michele Loi and the computer scientist Paul Olivier DeHaye have argued that "dominant tech platforms" should be viewed as basic social structures because of their pervasive influence over how we behave and interact with one another.

*A potential exception might be Apple, which is trying to position itself as the privacy-protecting big tech firm.

The best examples of this are the large social media platforms such as Facebook and Twitter. They affect how we communicate and interact with one another on a daily basis. Loi and DeHaye argue that, when viewed as basic social structures, these dominant tech platforms must be required to uphold the basic principles of social and political justice, which include protecting our fundamental freedoms such as freedom of speech and freedom of association. Platforms like Facebook even seem to be waking up to their responsibilities in this regard (although more cynical interpretations are available).[29] In any case the distinction between big tech and big government is, at least in some cases, a spurious one.

Finally, even if you reject this and think there are good grounds for distinguishing between the threats from big tech and big government, it is important to realize that just because it might be appropriate to take threats from government more seriously than threats from private enterprise, it doesn't mean we can discount or ignore the latter. Citizens still have an interest in ensuring that private enterprise does not unduly compromise their autonomy, and governments have a responsibility to prevent that from happening.

What Should We Do about It?

Now that we have a clearer sense of the nature and value of autonomy as well as the potential threats to which AI exposes it, we can turn to the question of what to do about these threats. We can start by noting that how you approach this question will depend on how you value autonomy. If you think autonomy is not particularly valuable or that it is only valuable to the extent that it enhances individual well-being, then you might be relatively sanguine about our current predicament. You might, for example, embrace the paternalistic ethos of hypernudging and hope that, through AI assistance, we can combat our biases and irrationalities and lead longer, healthier, and happier lives. In other words, you might be willing to countenance a post-freedom society because of the benefits it will bring.

But if you reject this and think that autonomy is an important, perhaps central value, you will want to do something to promote and protect it. There is an old saying widely but probably falsely attributed to Thomas Jefferson that states that "eternal vigilance is the price we pay for liberty." Whatever its provenance, this seems like a good principle to take with us as we march into the brave new world that is made possible by pervasive AI. Comprehensive legal regulations on the protection of personal data and privacy—such as the EU's General Data Protection Regulation—are consequently a good start in the fight to protect autonomy because they give citizens control over the fuel (data) that feeds the fire of AI. Having robust consent protocols that prevent hidden or unknown uses of data, along with transparency requirements and rights to control and delete data, are all valuable if we want to protect autonomy.

But they may not be enough by themselves. We may, ultimately, require specific legal rights and protections against the autonomy-undermining powers of AI. In this respect, the proposal from Brett Frischmann and Evan Selinger, for the recognition of two new fundamental rights and freedoms is worth taking seriously.[30] Frischmann and Selinger argue that in order to prevent AI and algorithmic decision-making systems from turning us into simple, unfree stimulus-response machines, we need to recognize both a freedom to be off (i.e., to not use technology and not be programmed or conditioned by smart machines) and a freedom from engineered determinism. Frischmann and Selinger recognize that we cannot be completely free from the influence and interference of others. We are necessarily dependent on each other and on our environments. But the dependence on AI is, they argue, different and should be treated differently from a legal and political perspective. Too much dependence on AI and we will corrode our capacity for reflective rationality and independent thought. Indeed, at its extreme, dependence on AI will obviate the need for autonomous choice. The AI will just do the work for us. The two new freedoms are designed to stop us from sliding down that slippery slope.

As they envisage it, these freedoms would entail a bundle of positive and negative rights. This bundle could include rights to be let alone and reject the use of dominant tech platforms—essential if we are to retain our capacity to resist manipulation and interference—as well as positive obligations on tech platforms and governments to bolster those capacities for reflective rationality and independent thought. The full charter of relevant rights and duties remains to be worked out, but one thing is obvious: public education about the risks that the tech poses to autonomy is essential. As Jamie Susskind notes in *Future Politics*, an active and informed citizenry is ultimately the best bulwark against the loss of liberty.[31]

Summing Up

So what can be said by way of conclusion? Nothing definitive, unfortunately. Autonomy is a relatively modern ideal.[32] It is cherished and protected in liberal democratic states but is constantly under siege. The rise of AI introduces new potential threats that are maybe not categorically different from the old kind but are different in their scope and scale. At the same time, AI could, if deployed wisely, be a boon to autonomy, enhancing our capacity for rational reflective choice. It is important to be vigilant and perhaps introduce new rights and legal protections against the misuse of AI to guarantee autonomy in the future. However we think about it, we must remember that autonomy is a complex, multidimensional ideal. It can be promoted and attacked in many different ways. If we wish to preserve it, it is important to think about it in a multidimensional way.

8 Algorithms in Government

This chapter is about the use of algorithmic decision tools by governments, specifically government agencies. Of course, *every* chapter in the book so far has (to some extent) touched upon government uses of AI and algorithmic decision tools. Consequently, you may wonder whether a chapter dedicated solely to the use of such tools by government agencies has anything new to add to the discussion. We think it does. The issues raised in previous chapters are relevant to how we assess the use of algorithmic decision tools by government agencies, but there are distinctive issues that arise from this use too, and it's worth giving them a fair hearing. It will help if we start with a clearer characterization of what those issues might be.

In *On the Wealth of Nations,* Adam Smith argues that specialization (or what he calls the "division of labor") is a key engine of economic growth. He illustrates this with the example of a pin factory, which makes pins consisting of a pointed metallic shaft and a flat metallic cap. He observes that "a workman not educated to this business" would struggle to make more than one of these pins per day "and certainly could not make twenty." But if the process of making the pins is broken down into a number of distinct processes, and if individual workers are trained and specialized in performing those distinct processes, a team of ten pin-makers can make "upwards of forty eight thousand pins in a day." And so, with specialization, you can dramatically increase the productivity of labor and, by extension, the "wealth of nations."[1] Specialization of this Smithian kind has long been embraced by the private sector. It is at the heart of most businesses and firms in existence today. This is also reflected in how the private sector makes use of technology. Where once it was teams of human workers who trained in specialized tasks, now it is teams of workers, machines, and algorithms that specialize in task performance. (We already caught wind of this shift in chapters 4 and 5.)

But the value of specialization is not just limited to productivity in the private sector. It's a general principle of social organization. As Thomas Malone notes in his fascinating book *Superminds*, the success of human civilization is largely attributable to our "collective intelligence," that is, how well we work together in groups to solve problems.[2] As the problems we confront get more complex, a good rule of thumb is to try to break them down into more manageable sub-problems and get specific individuals or organizations to specialize in the resolution of those sub-problems. They can bring their unique knowledge to bear on those problems.

This rule of thumb has been applied with vigor to the management of modern nation-states. The business of government has probably never been simple, but today it is a massively complex task. Governments create specialized agencies to solve all manner of societal problems. These include agencies dedicated to managing and regulating healthcare, the payment of welfare, finance, public and private transport, communications technologies, data gathering technologies, energy usage, environmental protection, food safety, drug safety, and so on. The swelling in both the number and size of these agencies is one of the hallmarks of the "administrative state" that has become the norm in liberal democratic regimes since the mid-twentieth century. These agencies are often created in response to particular crises and at moments of political convenience. Nevertheless, the advantages of specialization are often very real and tangible. No single politician or elected official could possibly govern an advanced, multicultural, industrialized nation without the help of specialized problem-solving agencies.

But this specialization creates a problem from a democratic perspective. The creation of specialized agencies changes the relationship between citizens and institutions of power. In particular, it risks *attenuating the legitimacy relationship* between them. What does this mean? In broad outline, the problem is this: specialized government agencies usually have the power to affect individual citizens in significant ways. They can deny rights and privileges, impose penalties and fines, and otherwise interfere with their capacity to live a flourishing life. To give an obvious example, a government agency that denies welfare payments to a person with no other source of income is doing something that could profoundly impact that person's capacity to live a good life. In liberal democratic states, the fundamental moral assumption is that power of this sort cannot be exercised unless it is *legitimate*.[3] What that means depends to some extent on who you talk to. Philosophers and political theorists have identified many different "legitimacy conditions" that may need to be satisfied in order for power to be exercised legitimately. The most obvious way to ensure legitimacy is to get the consent of the person who might be affected by the exercise of power. This is the route favored by private enterprises when getting customers to sign contracts (though whether they live up to that ideal is a matter that was touched upon in the last chapter). But government agencies can rarely rely on such consent to legitimize their power. Governments don't behave like merchants selling wares to passers-by. They can't feasibly transact with every citizen individually (not as a rule anyway). Instead, governments conduct "business" for their populations as collective wholes. But this obviously runs the risk of acting against the express wishes of *at least some* members of these populations. So they rely instead on giving citizens a meaningful say in shaping the institutions of power.

The most obvious way to ensure that citizens get a meaningful say in shaping the institutions of power is to either consult directly with citizens on the construction of those institutions (e.g., through constitutional ratification and amendment) or to get citizens to elect *representatives* who make laws and shape institutions of power on their behalf. The most pared-down version of parliamentary democracy is the clearest illustration of this method of legitimation. An individual citizen votes for a representative and then the representative votes on specific legislative proposals in the parliament. If the citizen does not like the way the representative votes, they can hold them to account at public meetings and clinics or use the power of the ballot box to vote them out at the next election.

In practice, however, the creation of specialized government agencies often compromises these lofty ideals. As these agencies are typically created by parliaments and not directly by citizens through plebiscite they come into existence at one step removed from the most powerful source of legitimation: the direct will of the people. This might be acceptable if we assume that (a) the directly elected representatives in the legislative assembly are exercising their power to create such agencies legitimately and (b) they retain control over how those administrative agencies function. But this may not always be the case. Often specialized agencies are created in such a way that they are directly insulated from the whims and vagaries of electoral politics. This is often intentional. They are designed so that they have some independence from government and so are not swayed by the same short-term, interest-group-guided concerns of elected officials.

A classic example of this is the setup for the typical central bank. Though the form varies, a central bank will usually have ultimate control over a country's financial system, acting as both a lender of last resort to private banks and the "printing press" that controls the money supply. From long and unfortunate experience, we have learned that these powers can be abused if exercised by elected governments, and so we now favor an institutional design that insulates central banks from direct government interference. But this, of course, means that central banks have significant and wide-ranging "unelected power."[4] This leads to many people questioning their legitimacy, particularly in times of financial crisis. People rail against the "technocrats" running these institutions and the power they have over our lives.

This example is, of course, just the tip of the iceberg. As specialized agencies proliferate, the problem of unelected power becomes more pervasive, and it can be exacerbated when the agencies are given considerable discretionary authority to design and enforce policies, and are allowed to outsource or subcontract their powers to others. The net result is a system of governance in which those who exercise power are further and further removed from the legitimizing source of that power: the citizens who are affected by it. The gains in problem-solving efficiency (assuming they are real) come at the expense of legitimacy. This is was what we meant when we earlier referred "attenuating the legitimacy relationship."

Of course, this is not a new problem. Governments have been dealing with it since the birth of the administrative state. They try to safeguard against it through various policies and legal doctrines. The names of these policies and doctrines vary from country to country, but broadly speaking we can identify three conditions that need to be satisfied if the creation of and exercise of power by specialized agencies is to be legally and politically justifiable.

1. **Sound policy rationale:** There must be some compelling public interest that is served by the agency and its use of power, and its policies and practices must be directed at that public interest.
2. **Appropriate delegation of power:** The agency must have its powers conferred upon it in a legitimate way by some legally legitimate authority. Usually this will be through a piece of legislation that specifies what the agency is supposed to do and how it is supposed to do it. The legislation may give discretion to the agency, including the power to subdelegate to third parties (including private companies), but usually this is tolerable only if there are clear limits to the discretion.

3. **Compliance with principles of natural justice/fair procedures:** The agency must comply with widely recognized principles of natural justice, due process, or fair procedures (different terms are used in different jurisdictions) when exercising its powers against citizens. This typically means that the agency should have some plausible reason or explanation for why it exercised its powers in a particular way, that it not exercise its power in a discriminatory and unfair way, and that the person affected has the right to be heard and the right to appeal (or review) a decision to an impartial tribunal. (See box 2.1 in chapter 2 on the difference between "appeal" and "review.")

The practical meaning of each of these conditions can be quite technical and complex in particular legal traditions. We'll encounter some of this complexity below, but for now, all that matters is that the conditions are understood at a general level.

This gives us everything we need to understand the problem that might arise from the use of algorithmic decision tools by government agencies. The *general* concern is that the proliferation of specialized agencies attenuates the legitimacy of power; the *specific* concern is that the use of algorithmic tools by those agencies might further exacerbate this attenuation of legitimacy. Although there are safeguards in place to deal with the attenuation of legitimacy at a general level, we might wonder whether algorithmic tools pose novel or unexpected problems when it comes to the attenuation of legitimacy that cannot adequately be met with those traditional safeguards. In other words—and to continue the theme from earlier chapters—we might ask, is there something different when it comes to the use of this technology?

Case Study 1: The Go Safe Automatic Speed Cameras

Before we get too mired in the legal and philosophical debate, let's consider a case study. This case study illustrates some of the ways in which the use of power by specialized agencies can go wrong, particularly when a technological aid is used to assist in the exercise of that power. The case study comes from Ireland, and it concerns the use of automatic speed cameras to prosecute speeding offenses.[5]

Ireland, like many countries, treats speeding as a minor offense. If you are ever caught speeding on an Irish road, you will more than likely receive a fixed penalty (or charge) notice in the mail. The penalty will consist of a fine, the application of "penalty points" to your driver's license, or both. If you clock up too many penalty points, you might be banned from driving for a period of time. In most cases, when people receive these fixed penalty notices, they pay the fine immediately and think no more about it. If they fail to pay the fine by a due date, they might be summonsed to court and face a slightly more severe criminal prosecution. Under the relevant legislation in Ireland (the Road Traffic Act 2010, as amended) the Irish police force is given the authority to detect and prosecute speeding offenses. There is, however, a section of that act, section 81(7), that allows for this power to be delegated to a third party (e.g., a private company) as long as this is done via a written agreement entered into with the Irish Minister for Justice.

In 2009, the Irish police force, with the approval of the Irish government, decided to outsource the detection and prosecution of speeding offenses to a private company called Go Safe. They did so under a contract with the Minister for Justice that gave Go Safe the right to do this until 2015, with the option for renewal then. The company operated a fleet of vans with automatic speed cameras. These vans were placed in strategic locations for a number of hours. These locations had to be approved by high-ranking members of the Irish police force. While *in situ* the cameras in the vans would automatically detect violations of the local speed limit and take photographs of the offending vehicles. Subsequently, fixed penalty notices would be issued to the registered owners of those vehicles. Much of this process would take place automatically, with some minimal oversight by human operators. If the registered owner refused to pay the fixed penalty, they could be summonsed to court, and Go Safe workers would give evidence in court to support their prosecution.

On the face of it, this would appear to be a textbook example of how to effectively use technology to administer the business of the state. There is a clear and obvious public interest at stake—speeding is a contributing factor to road deaths and reducing it protects public safety.[6] Automatic speed cameras can accurately detect speeding violations without the need for constant human supervision and input. The mere presence of those cameras, or the suspicion of their presence, has a deterrent effect and prompts drivers to change their behavior. Using them saves precious policing resources and enables the efficient administration of justice. What's more, the way in which the power to detect and prosecute was given to the Go Safe company does not appear to raise any obvious red flags when it comes to the attenuation of legitimacy. There was a piece of legislation that authorized the outsourcing of this power to a third party; there were statutorily prescribed limits to how much power they could have; and there had to be a written agreement spelling out the terms and conditions of their service.

But things didn't work out so smoothly in practice. Road traffic offenses are notoriously complex. There are a number of technical protocols that need to be followed to legally prove their occurrence. It is not uncommon for defense lawyers to use this technical complexity to the advantage of their clients, pointing out how certain protocols were not followed thus rendering a prosecution void. Once defense lawyers got to work on figuring out the flaws in the Go Safe system, things quickly turned sour. According to one report,[7] over 1,400 attempted prosecutions of speeding offenses using the Go Safe system were thrown out of court. A variety of grounds were given for these dismissals. One of the biggest problems was that Go Safe could not prove whether fixed penalty notices had been received by potential offenders or that registered owners were in fact driving their vehicles at the relevant time. A whistleblower from the company complained that the company demanded that its workers record potential speeding violations even when they couldn't set up the camera equipment appropriately and so would be running the risk of false positive identifications.[8] Judges complained that when officials from the company gave evidence in court they could not explain how the camera system worked, how far above the speed limit someone had to drive before being issued a fixed penalty notice, or how the fixed penalty notices were generated and sent to potential offenders. They also lamented the failure to provide an adequate chain of evidence and to prove that they had the authority to give evidence in court. In short, the practical implementation of the Go Safe system was a shambles, so much so that one Irish judge referred to it as an "abject failure."[9]

Some of the problems with the Go Safe system have since been addressed,[10] and the company's contract was renewed in 2015.[11] Nevertheless, the Irish experience with the Go Safe system provides us with a cautionary tale. Automatic speed cameras are not an advanced technology. They use simple radar reflection to calculate the speed of oncoming vehicles, determine whether this exceeds the local speed limit, and then take a photograph. They are nowhere near as sophisticated as some of the algorithmic tools discussed in this book. They don't make complex predictions or judgments. They don't depend on esoteric programming techniques or advanced artificial intelligence. Even still, their use in the administration of justice caused untold practical headaches. A new technological system had to integrate itself into an old legal governance system—like a software upgrade on an old computer. The upgrade did not go well. Its problems were compounded by the fact that it involved a contractual relationship between a government agency and a private corporation. Simple errors and omissions were made when implementing the system. People representing the company could not adequately explain how their systems worked. Sometimes incentives were misaligned and corners were cut. The result was a failure to meet a key public policy objective and a failure to legitimately delegate power.

If all these things can happen with a relatively straightforward technology, we would be well advised to be on our guard with a more complex one.

Does the Use of Algorithmic Decision Tools Pose a Threat to Legitimacy?

Granted that we need to be careful about the use of algorithmic decision tools by government agencies, is there any reason to think that special caution is required? Or, to go even further, could it be that the use of algorithmic decision tools poses such serious risks to the legitimate use of power that there should be a general presumption against their use?

Several commentators have expressed skepticism about the need for such hyper-caution.[12] They point out that the administrative state has long battled with criticisms concerning the legitimate use of power, and it has survived these criticisms despite increased specialization and increased use of public-private partnerships in how it carries out its key functions. Even the fiasco of the Go Safe system in Ireland did not lead to any major doubts about the use of technology by government agencies or the legitimacy of public-private partnerships. It just led to changes in policy and practice. It seems likely that a similar pattern will be followed when it comes to the use of algorithmic tools by government agencies. Nevertheless, it is worth entertaining a hyper-cautious stance, if only to see what might be wrong with it.

To do this it helps if we clarify exactly how algorithmic tools can be and might be used by government agencies. Government agencies perform two main functions: (1) they create policies and rules (if they have been given the discretionary authority to do so) and (2) they implement and enforce policies and rules (either the ones they themselves have created or those that have been stipulated for them by other elected authorities).

Algorithmic tools can help with both tasks. Examples have been given in previous chapters. Algorithmic tools are already being used by government agencies to efficiently manage the implementation and enforcement of rules, for example, in predictive policing software that provides guidance on how to deploy policing resources, and to identify and repair flaws in existing policies or rule-making frameworks, for example, in smart energy grid systems or traffic signaling systems. Sometimes algorithmic tools will be used as mere aids or supplements to human decision-making within the relevant government agencies. At other times algorithmic tools will operate autonomously with minimal human supervision and interference. In their discussion of the issue, Cary Coglianese and David Lehr refer to the possible autonomous uses of algorithmic decision tools as "adjudication by algorithm" and "regulating by robot," respectively.[13] These names are chosen for their alliterative appeal, not for their descriptive accuracy. For example, as the authors point out, what they call "regulation by robot" may not involve a robot; it may just involve an algorithmic system making rules without direct input from a human controller.

The merely assistive and supplementary use of algorithmic decision tools seems to raise relatively few red flags when it comes to the legitimate use of power. If the algorithmic tools are being used *as* tools, then human decision makers retain the actual power, and any issues we might have with how they exercise that power are ones that we are ultimately familiar with. The existing legal and regulatory framework is designed to work with human decision makers. By contrast, genuinely "autonomous" uses of algorithmic decision tools have the potential to raise more red flags.* Depending on how much autonomy is given to a tool, there is a risk that humans will no longer be the ones in charge (i.e., in meaningful/effective control). Questions might then arise as to whether the existing legal and regulatory framework is fit for purpose. It is supposed to be government by *humans* for humans, not government by *machines* for humans.

There is, however, a danger that we get carried away when it comes to assessing the potential threat that autonomous algorithms pose to the legitimacy of government. In this respect, we are easily seduced by fictional

*In terms of chapter 5, assistive and supplementary tools don't pose the control problem. On the other hand, fully automated decisions (and subdecisions) do pose this problem.

motifs of machines taking over the reins of power. We are a long way from that with current technologies. A more sober analysis is needed, taking into consideration the three conditions for the legitimate use of power by government agencies that were outlined above. When assessed in light of these three conditions, does the use of autonomous algorithmic decision tools really pose a significant threat?

Well, we can quickly set aside the first condition. Whether there is a sound public policy rationale for the use of a decision tool will vary from case to case, but we can easily imagine that there is often going to be one. For example, we could create algorithmic adjudicators that scrape through data on financial transactions and automatically impose sanctions on entities that breach financial rules. These algorithmic adjudicators could very credibly do a better job than human adjudicators. Financial markets are already suffused with trading bots and algorithms, executing thousands of trades in the blink of an eye. No human adjudicator can keep on top of this. An algorithmic adjudicator might be just what we need to ensure the smooth functioning of the regulatory system.[14]

That leaves us with the other two conditions: (1) whether power has been legitimately delegated to the algorithmic tool and (2) whether the use of the algorithmic tool complies with principles of natural justice/fair procedures. The issue of legitimate delegation is tricky. As Coglianese and Lehr point out, there are a number of different ways in which power could be delegated to an autonomous algorithm.[15] The first and most straightforward would be if a piece of legislation explicitly provides for the use of the tool. Imagine, for example, a road traffic act that stated explicitly that the police force is entitled to use "algorithmic adjudicators" to determine whether someone has breached a speed limit and to automatically impose penalties if they do. As long as the legislative provision is clear about this—and it is supported by some intelligible public policy rationale—the delegation of power to the algorithm would be legally uncontroversial. That's not to say that it would be wise from a political or public policy perspective (a point to which we return); it is just to say that it doesn't create special problems from a legal perspective. This is the normal and most legally appropriate way to delegate power. In Australia, section 495A of the Migration Act authorizes the minister of immigration to use computer programs when making certain decisions, and decisions by the computer program are then taken to be decisions by the minister.

The second way in which power could be delegated to the algorithmic tool is through discretionary subdelegation by someone within a government agency. This is not an uncommon practice. Elected governments create specialized agencies to manage and regulate key sectors of society because they realize that they themselves lack the requisite knowledge and expertise. Consequently, they have to give the staff within those agencies some discretion as to how best to create and implement rules and policies. So, for example, there might be a road traffic act that gives the police force the power to use "whatever means they see fit" to enforce the speeding laws. An official working within the police force might read about a new "algorithmic adjudication" system that would allow them to enforce the speeding laws more efficiently. Following appropriate consultation and procurement, they might decide to use this tool to exercise the discretionary power that has been delegated to them by the road traffic act.

But this discretionary subdelegation to an algorithm might be legally problematic. Questions would necessarily arise as to how to construe the precise scope of the discretionary power of subdelegation. A provision that allows an agency to use "whatever means they see fit" might be deemed too vague and open-ended to legitimize that form of delegation. Surely the police couldn't use "any means"? Imagine if they started using smart AI-based landmines that detonated underneath cars that exceeded the speed limit by 0.1 of a km/h? That would surely be shut down, largely because it would not comply with other principles underlying the legitimate use of power (e.g., principles of fair and proportionate punishment). This is a silly example, but it illustrates an important point. Even where the discretionary power given to an agency under a statute seems quite broad, there must be some limits to it. Discretionary powers that are phrased in terms of using "appropriate means" or "proportionate means" will raise similar issues of scope uncertainty.

Writing from the US perspective, Coglianese and Lehr argue that, although the subdelegation of power to an algorithmic tool will raise questions as to how best to interpret the power of subdelegation under the relevant statute, there is "in principle" no reason why power could not be subdelegated to an algorithm.[16] They make two points in support of this argument. First, they argue that government agencies already unproblematically make use of measurement devices that perform the kinds of functions that could be subdelegated to an algorithm. No one questions their right to do so on grounds of illegitimate delegation of power. Second, they argue that given the way in which algorithmic tools currently operate (they focus specifically on machine learning algorithms) humans will always retain some control over them, either by determining the objectives/goals they are supposed to meet or determining how and when they are to be used. There will never be a complete subdelegation of power to an algorithm.

But we might question Coglianese's and Lehr's sanguine outlook. In chapter 5, we considered the issue of control in relation to the use of algorithmic decision tools and spoke about the risk of "automation bias." As we argued, there is a very real risk that when people rely on algorithmic systems they might defer to them excessively and lose meaningful control over their outputs. This could happen in government agencies too. Government agencies that rely on autonomous algorithmic tools may adopt an uncritically deferential attitude toward them. They may, in principle, retain some ultimate control, but in practice it is the algorithm that exercises the power. The British sketch comedy show *Little Britain* satirized this problem in one of its recurring sketches involving a receptionist who could never answer a customer's query because "the computer says 'no.'" A similar attitude might creep into government agencies. The human officials may not be inclined to wrestle control back from algorithmic tools, not because the tools are smarter or more powerful than they are, but because habit and convenience make them unwilling to do so. Looking at this topic from the perspective of UK administrative law, Marion Oswald argues that this kind of deference to the machine would be legally problematic no matter whether the algorithm was developed by the agency themselves or by some third party.

> A public body whose staff come to rely unthinkingly upon an algorithmic result in the exercise of discretionary power could be illegally "fettering its discretion" to an internal "home-grown" algorithm, or be regarded as delegating decision-making illegally to an externally developed or externally run algorithm, or having predetermined its decision by surrendering its judgment.[17]

Before this happens, we might like to take some corrective action and ensure that we don't allow government agencies to fetter their discretionary power to such an extent. We might like to create a new norm whereby any delegation of power to an autonomous algorithm has to be done on the basis of an explicit legislative provision, not through discretionary subdelegation.

Finally, what about natural justice and fair procedures? Is there any risk that the use of algorithmic tools poses special problems in this regard? The concept of natural justice or fair procedures overlaps significantly with topics discussed in previous chapters. A fair procedure, broadly construed, is one that is relatively impartial,* takes into consideration the interests of the parties affected, gives them a right to be heard or consulted, provides reasons for the decisions made, and gives the affected person a right of appeal in the event that they still feel aggrieved. It is certainly possible that the use of an algorithmic tool would violate these requirements. The problems of bias and transparency, discussed in earlier chapters, would be relevant in reaching such a conclusion. If an algorithmic adjudicator makes decisions in a systematically biased and opaque way, then it may very well fall foul of the fair procedure requirement. But, as was pointed out in those earlier chapters, there are some difficult tradeoffs when it comes to eliminating bias, and there are ways to ensure that algorithmic decisions are explainable. So even though an algorithmic adjudicator could fall foul of the fair procedure requirement, it doesn't have to. There is no categorical reason to oppose the use of algorithmic tools by government agencies on these grounds.

There is another point worth making here too. Although there is an idealized conception of what a fair procedure should look like, many legal systems do not insist that every decision made by a government agency meet this ideal. Some corners can be cut in the interests of efficiency and cost-effectiveness. This makes sense. If an agency had to hold an impartial tribunal and provide detailed explanations for every decision they made, the actual day-to-day business of that agency would grind to a halt. Courts usually accept this and adopt "balancing tests" when figuring out how close to the ideal of a fair procedure any particular decision-making process must get. For example, US courts focus on three factors: how the decision impacts the affected party, the potential cost to that individual if a wrongful decision was made, and the net gain (if any) from introducing additional procedural safeguards to protect against that potential cost. They then weigh these three factors and decide whether a particular decision-making process is acceptable or needs to be reformed to pass muster.[18] Given this sensitivity to cost-effectiveness, it's quite likely that many uses of algorithmic tools will be deemed legally legitimate even if their use prevents a decision-making procedure from living up to the ideals of natural justice. After all, one of the reasons why government agencies might be tempted to use this

* We say "relatively" because absolute impartiality is impossible.

technology is to help them manage complex systems in a cost-effective manner. Think back to the earlier example of an algorithmic adjudicator enforcing financial regulation against algorithmic trading bots.

None of this is to say that the use of algorithmic decision tools by government agencies is always going to be a good idea. A technological aid might be a good idea in principle but a bad idea in practice. Furthermore, even a tool that does enable the more effective management of social systems and is technically legally legitimate might not be *perceived* as legitimate. Public administration is as much about good public relations as it is about legal technicalities and economic efficiencies. Any government considering the use of such a system would be well advised to consult widely on its introduction, listen to the concerns of key stakeholders, and constantly review the practical workings of the system once it is up and running. These are standard practices in government agencies the world over anyway, but it is important that these practices are maintained and emphasized.

It is also not to say that the use of algorithmic decision tools throws up no new challenges for the legitimacy of public administration. They do. Algorithmic tools work on the basis of a precise quantified logic. Human decision makers often work more by qualitative reasoning and intuition. This means that if some authority is delegated to such tools, there is going to be a need to translate what was once a qualitative decision-making process into an explicitly quantitative one. This translation process might throw up some new issues. It is impossible to create a flawless decision-making process. There is always some risk of error. An automated speed camera system, for example, might sometimes (even if only rarely) fail to record a car that is traveling over the speed limit or mistakenly record a car that was not. The former would be a false negative error; the latter would be a false positive. We live with the risk of both errors all the time, but we often don't think about them in explicitly quantified terms. In other words, we don't explicitly decide that we are okay with a system that makes false positive (or false negative) errors 5 percent or 10 percent of the time. We often live with the illusion that we are aiming for perfection. The use of algorithmic decision tools will force us to discard this illusion. Although some government agencies are already comfortable with making explicit choices about error rate, others may not be. The fact that they might *have* to may also create a crisis of perceived legitimacy, as the public has to confront the quantitative realities of risk. If a government department is going to use an autonomous decision tool, it should be able to support this choice by making publicly available evaluations showing that the tool's decisions are equivalent to, or better than, the decisions of the relevant human staff.* But this requires delicate handling.

There is another, more philosophical, problem with the quantitative logic of algorithmic decision tools. It has to do with how these tools treat individual human beings (see the previous chapter). As noted earlier, in liberal democratic states, the legitimacy of public power is founded on respect for individual citizens. They are autonomous, dignified beings whose lives must be taken into consideration. To paraphrase the philosopher Immanuel Kant, they must be treated as whole, integrated persons—as ends in themselves—rather than as means. A long-standing concern with the specialization and bureaucratization of public administration is that it fails to live up to this Kantian ideal. The dignity of the individual is undermined by complex, mechanized management. Individuals are cogs in the machinery of the state. They are confronted by labyrinthine processes and nameless officials. They are bundles of statistics, not fully-rounded characters. They are "cases" to be managed, not persons to be respected. Although this dehumanization concern is a long-standing one, it is a concern that could be exacerbated by the use of algorithmic decision tools. These tools necessarily reduce persons to bundles of data. They have to quantify and disaggregate people's lives into mathematically analyzable datasets. They don't "see" people; they see numbers. Give this necessity, special safeguards may need to be put in place to maintain the human touch and ensure that the dignity of the citizen is respected.

Case Study 2: The Allegheny County Family Screening Tool

Let's now consider another case study in the use of technology by government agencies. Unlike the Irish case study, this one involves a more complex algorithmic decision tool. The case study concerns the Allegheny County Family Screening Tool (AFST for short).[19] We mentioned it briefly in chapter 5 already. We return to it here to see whether it holds any lessons for the legitimate use of power by government agencies.

---

*We discussed evaluation methods in chapter 1.

The AFST is, as the name suggests, a screening tool used to identify children who may be at risk of abuse and neglect. It was created by a group of academics, led by Rhema Vaithianathan and Emily Putnam-Hornstein.[20] The group was originally commissioned by the New Zealand Ministry of Social Development to create a predictive risk modeling tool that could sort through information about how families interact with public services and criminal justice systems to predict which children were at the most risk of abuse or neglect. The tool is supposed to use this information to generate a risk score for each child that can then be used by child protection workers to investigate and prevent cases of abuse and neglect. In this respect, the tool is not dissimilar to predictive policing tools that generate heat maps to assist police departments with the efficient distribution of policing resources. Indeed, the visual display used by the family screening tool is quite similar to that used in predictive policing heat maps. It adopts a traffic light warning system that identifies high-risk cases with a red light and low-risk cases with a green light.

The team led by Vaithianathan and Putnam-Hornstein eventually came up with the predictive model that relied on 132 different variables—including information about the mother's age, whether the child was born to a single parent family, mental health history, criminal record, and so on—to generate its risk scores. They claimed that the system was reasonably accurate at predicting abuse and neglect. But they ran into problems when trying to implement it in New Zealand.

The role of the government in child protection has always been politically contentious. At the extreme, child protection law entitles government agents to take children away from their lawful parents. This is often an upsetting experience for both the parents and the children. It is usually a last step. In most countries, child protection caseworkers only escalate to this step when other interventions have failed, but historical experience has made some people very suspicious of the process. There is often the belief that government agents unfairly target poor parents from minority backgrounds or those with unconventional lifestyles, and there is a sordid history of children from indigenous minority ethnic families (e.g., New Zealand, Australia, and North America) being taken from their families at disproportionate rates.[21] It is unsurprising, then, that the New Zealand government halted the experimental use of Vaithianathan and Putnam-Hornstein's tool in 2015.[22] (Strictly speaking, the experiment was aborted because the minister didn't want to use children as guinea pigs rather than because she acknowledged that the tool would be discriminatory, but the issues are often difficult to disentangle.[23])

The Allegheny County Department of Human Services (DHS)—based in Pennsylvania, however, did award a contract to the group to create the AFST. They have used the system since August 2016.[24] It is clear that, in designing and implementing the AFST, both the academic team and the Allegheny County DHS were very sensitive to concerns that local citizens might have about the tool. They held a number of meetings with key stakeholders in the county to determine how best to create and implement the system. They also hired an outside, independent team to conduct an ethical analysis of the AFST. This team concluded that use of the tool was ethically appropriate because it was more accurate than preexisting systems for screening cases.[25] Furthermore, the group behind the AFST have tried to ensure that the system operates in a highly transparent way, releasing information about the variables used and providing detailed and updated FAQs to the public about how it works.[26] This has led to some public praise. For example, writing in the *New York Times*, the journalist Dan Hurley was largely positive about the impact of the AFST and considered it an important development in the fight against child abuse and neglect.

But others have been more critical. The political scientist Virginia Eubanks, in her book, *Automating Inequality*, argues that the AFST still unfairly targets children from poor and minority families, and unfairly correlates poverty and use of public services with an increased risk of abuse.[27] She has also argued that the creators of the system have not been transparent enough in releasing information about how the AFST works. They have released details about the variables they use but not the weights they attach to these variables. She also argues that the system may have an unacceptably high error rate, noting that officials at the Allegheny County DHS state that 30 percent of cases flagged by the AFST are ultimately thrown out as baseless,[28] but further noting that this cannot be properly assessed until information is released about the error rate.

The debate about the AFST perfectly illustrates the themes of this chapter. The AFST was created to enable a government agency to more accurately and more efficiently go about its business. But the way in which it operates creates concerns about the legitimate use of power. This is true even though many of the steps to ensuring legitimacy were followed. Critics still worry about the opacity and complexity of the system: they worry that it may be unfair or biased; they want to be more involved with its implementation; and they demand greater transparency and openness when they are. This is a cycle that never really ends. The legitimacy of power can always be called into question. It is incumbent on both citizens and those working in government agencies to be willing to scrutinize how power is wielded on an ongoing basis to ensure its legitimacy. The use of algorithmic decision tools adds a new technological flavor to this perpetual dynamic.

There is another lesson to be learned from the AFST case study. One of the major criticisms that Virginia Eubanks launches against the AFST and other systems like it is that they are examples of technological solutionism in practice.[29] We now have sophisticated, algorithmic risk-prediction tools. We know they are more accurate than human decision makers in certain cases, so we look around for problems that they could be used to solve. But we don't think it all the way through. We are so eager to find some application that we don't think about the side effects or indirect consequences of that application. Nor do we consider that there may be *other* problems—perhaps not so easily solved by risk-prediction software—that are more deserving of our attention. For instance, Eubanks worries that the AFST intervenes too late in the game, that is, when a child has called looking for help. At this stage in the game, the systematic biases and structural inequalities in society have taken their toll on the affected families. This is something that the independent ethical reviewers of the AFST themselves noted, observing that predictive risk modeling can be founded upon, and consequently help to reinforce, existing racial biases.[30] This is a point we made in chapter 3.

It is unrealistic to expect a tool like the AFST to completely correct for such structural inequalities, given its intended use. It has to work off a social reality in which members of poorer groups and racial minorities do suffer from disadvantages that may very well place them at a higher risk of some negative outcome. But in that case, the use of the tool may widen the perceived legitimacy gap, at least when it comes to members of those populations. Perhaps we should use this realization as an opportunity to reflect on how we want government agencies to use algorithmic decision tools. Perhaps there are other problems they could help us solve that might then reduce the legitimacy gap for those groups. This will obviously require more creative, outside-the-box thinking about how best to use algorithmic decision tools.

A Necessity for Algorithmic Decision Tools?

We started this chapter by noting that the proliferation of specialized government agencies has advantages when it comes managing social problems, but that this advantage comes with the risk of attenuated legitimacy. The use of algorithmic decision tools by those agencies must be interpreted and understood in light of that long-standing dynamic. Fortunately, we have developed political and legal safeguards to protect against the attenuation of legitimacy, and these can be used to evaluate and constrain the public use of algorithmic decision tools. This means that although there is no insuperable legal or political obstacle to the use of algorithmic decision tools, they should be deployed carefully and thoughtfully, and their use should be subject to public scrutiny and oversight.

In making this argument we have suggested, at various points, that the use of algorithmic decision tools could be a boon to government agencies. As the world becomes more complex and as algorithmic tools are used by private individuals and corporations, the use of similar tools by government agencies may graduate from being practically desirable to being practically necessary. It may be the only way to keep up. On the face of it, though, it seems like this is an argument that can only be made on case-by-case basis. Is there anything more general that can be said in favor of the use of algorithmic decision tools?

Perhaps. Consider the lessons to be learned from the history of societal collapse. In his influential book, *The Collapse of Complex Societies,* the archaeologist Joseph Tainter tries to provide a general explanation for the collapse of complex societies.[31] He looks at all the famous examples: the Egyptian Old Kingdom, the Hittite Empire, the Western Roman Empire, Lowland Classic Maya, and more. Although some archaeologists and historians dispute whether these societies actually collapsed—some argue that they simply adapted and changed[32]—it seems relatively uncontroversial to say that they underwent some decline in their social complexity (i.e., their administrative centers were dissolved; they abandoned settlements and discarded cultural and institutional artifacts). Tainter wonders whether there is some common cause of these declines.

After reviewing and dismissing several of the most popular explanations, Tainter comes up with his own explanation for societal collapse. His explanation consists of four key propositions. The first proposition is that complex societies are problem-solving machines. In other words, they sustain themselves by addressing the physical and psychological needs of their members. If they don't do this, they lose their legitimacy and ultimately collapse (or undergo significant decline). The second proposition is that in order to solve problems, societies have to capture and expend energy. Classically, societies did this through foraging and farming; nowadays, we do it through burning fossil fuels and exploiting other energy sources. The general rule of thumb is that, in order to solve more problems and sustain greater complexity, societies have to increase their energy expenditure (or at a minimum make more efficient use of the energy available to them). This leads to the third proposition—that the survival of complex societies hinges on a basic cost-benefit equation. If the benefits of societal problem-solving outweigh the costs, then the society will continue to survive; if they do not, then the society is in trouble. The fourth proposition states that an increased investment in social complexity (e.g., an increased investment in specialized administrative agencies) results in more benefits, but only up to a point. The increased complexity comes with an increased energy cost so that, eventually, the marginal cost of increased complexity will outweigh the marginal benefit. If this happens, and if the inequality is of a sufficient magnitude, the society will collapse. This problem of declining marginal returns is the core of Tainter's explanation. He argues that the complex societies that have collapsed throughout history have all confronted the basic problem of declining marginal benefits and increasing marginal costs.

Tainter gives several illustrations of how societies have had to deal with this problem. Some of the best documented examples have to do with bureaucratic power and control. Tainter shows how many government agencies undergo administrative bloat and mission creep. These agencies, often central to societal problem-solving, proliferate and grow in size to cope with new challenges. This is initially beneficial, as it allows the society to address more complex problems, but it ultimately results in declining returns as costly administrative staff have to be hired to manage the organizations themselves. There are ever more of these administrators compared to relatively fewer frontline caseworkers who actually deal with the problems those organizations are designed to address.

Tainter's theory prompts reflection on the role of technology in preventing societal collapse. Perhaps one way to solve problems efficiently without dramatically increasing the cost of administrative bloat is to double down on the use of algorithmic decision tools and other forms of AI. According to some of its advocates, a significant breakthrough in AI could be the *deus ex machina* we need to solve our growing social problems. Instead of relying on imperfect, squabbling human intelligences to manage our social problems, we could rely on a more perfected artificial form. This is an argument that Miles Brundage makes explicit in his paper on the case for "conditional optimism" about AI in global governance.[33] This doubling down on algorithmic decision tools and AI might come at the cost of some perceived legitimacy, but maybe that is the price we as citizens have to pay to maintain our complex social order. It is a provocative and disturbing suggestion, and may mean that we should be more open to the widespread use of algorithmic decision tools. Their legitimacy might be greater than we'd like to think.

Summing Up

The use of specialized agencies and specialized tools is essential to the modern business of government. You cannot govern a complex, industrial society without relying on expert agencies with specialized problem-solving skills. These agencies make judgments and set policies that have a significant impact on the lives of citizens and must exercise that power legitimately. The problem is that the increasing diversity and organizational depth of such agencies attenuates their legitimacy. The use of autonomous algorithmic decision tools can exacerbate this problem. Nevertheless, we shouldn't *overstate* the problem. Many countries have developed legal and political doctrines that can determine when and whether the exercise of power by a government agency is legitimate. There is no strong reason to think that these doctrines are no longer fit for purpose. We may just need to modify them and err on the side of caution when it comes to ensuring that power is legitimately delegated to algorithmic decision tools and that their use does not undermine the need for fair procedures. One way or another, AI will be increasingly indispensable if our ever more complex societies are to be prevented from collapsing under their own weight.

9 Employment

Anyone reading this book will have seen confronting news stories about robots replacing the human workforce and plunging millions into joblessness and poverty. We're so used to these warnings that we probably don't stop to think enough about their logic and take "the end of work" as something of a fait accompli. It's true that for the first time in history the advent of advanced machine learning and mobile robotics will intrude upon professional and managerial work and not simply (as has so far been the case) low-skilled "blue-collar" work. But the fact remains: there is no evidence supporting predictions of wholesale unemployment in the short to medium term caused by AI.

The science fiction writer William Gibson was fond of saying, "The future is already here—it's just not very evenly distributed." And sure enough, behind all the dire scenarios of mass worker displacement lie predictions about the uneven distribution of the profits from the AI revolution. Tech writers making gloomy predictions can readily find support in the grim economic statistics from recent decades.[1] Wages of ordinary workers have been stagnant or falling in many countries since the 1970s. The wages of production and nonsupervisory workers in the United States reached their zenith in 1973. In real terms, such workers now earn 87 percent of what they earned then.[2] At the same time, productivity has continued to rise. In the United States, as in many other countries, the lockstep relationship between productivity and the earnings of workers broke spectacularly in the mid-1970s and shows no sign of reviving. As the share of national income that goes to wage and salary earners continues to dwindle, the seemingly inexorable rise of inequality has prompted calls for new mechanisms to redistribute wealth away from the superrich.[3]

But fears about automation are as old as automation, and there is little agreement about how much of the recent bad economic news can be laid at the door of artificial intelligence and robotics. On the one hand, automation has often paid significant dividends, including increased productivity, the democratization of goods and services previously only available to the wealthy, and the increased autonomy and social mobility made possible by labor-saving appliances.[4] Authors such as Erik Brynjolfsson and Andrew McAfee have argued that AI will bring about a new industrial revolution, promising great advantages to those who manage to exploit the boom in new technologies.[5] On the other hand, it's worth remembering that most people in the first industrial revolution didn't live long enough to enjoy its benefits. The first industrial revolution spanned the eighteenth and much of the nineteenth centuries, but for most of this time it provided little benefit to ordinary workers who suffered displacement and squalid living conditions. The average height of English men decreased by 1.6 centimeters per decade over the eighteenth century.[6] For all the exciting technological advances, real wages simply couldn't keep up with the price of food. Despite rising output per worker, the purchasing power of their wages stagnated. Indeed, it wasn't until the mid-nineteenth century that real wages finally began to keep pace with productivity.[7]

What Do We Know about the Future of Work?

Debate about the effects of automation on human society has always come in two flavors. Optimists argue that automation increases economic growth, which benefits society overall. Some even suggest that it will eventually free humanity from the shackles of having to produce the necessities of life. Amongst the optimists, we find one of the most famous economists of all time—John Maynard Keynes. In his 1930 essay, "Economic Possibilities for our Grandchildren," he argued that automation and industrialization more generally would solve "the economic problem" of "the struggle for subsistence."[8] At the other end of the spectrum, the economist Jeremy Rifkin's book, *The End of Work*, paints a very different picture of mass technological unemployment and hardship.[9] These pronouncements about the effects of automation are not new. They mirror eighteenth century debates about the likely result of the first industrial revolution between writers such as David Ricardo and William Mildmay. So are we any closer to being able to predict what the current technological revolution will bring?

Much recent debate about the effects of AI and robotics on jobs has been in response to a 2013 study by Carl Frey and Michael Osborne at the University of Oxford. They found that 47 percent of total US employment had a high probability of computerization from machine learning and mobile robotics.[10] This result has been controversial, particularly as subsequent studies produced radically different results. It was pretty clear from the start that the size of predicted effects on the future of work depended strongly on the authors' methodological choices. Frey and Osborne modeled whole jobs—that is, job *categories*. Strictly speaking they found that of 702 precisely defined job categories, 47 percent were ripe for near-term automation. But notice what this means: if some component of a job was deemed to be at risk of automation, the *whole* job was assumed to be at risk. This seems counterintuitive. The fact that the typical bank clerk's job description today might only faintly resemble what it was in, say, 1980, *doesn't* mean there are no more bank clerks. More generally, automation has a way of redrawing boundaries around traditional job categories. Maybe the typical secretary or personal assistant no longer needs to come equipped with Pitman shorthand, but they might now be expected to perform functions that weren't historically typical, for example, monitoring a company's social media activity or attending more high-level meetings than was usual in earlier days. Understandably, a number of studies conducted after Frey's and Osborne's have focused less on jobs and more on *tasks*. A 2016 OECD task-based study found that only 10 percent of all jobs in the United Kingdom and 9 percent in the United States were fully automatable.[11] Other methodological factors that influence the results of such studies include how they interpret automation and computerization; timescale; judgments about likely scientific advances; and various social, economic, and demographic assumptions (e.g., about levels of migration, the global economy, etc.).[12]

This last set of assumptions can't just be swept under the rug either. Indeed, what probably makes all prognostication here a little amateurish is the fact that the shape any future workforce is likely to take is going to depend a lot on whether the jobs (or tasks) that remain are going to force a reappraisal of skills that have traditionally been undervalued. For instance, we know that women are overrepresented in the caring professions—in jobs like nursing, teaching, and counseling.[13] These jobs require empathy and people skills, what's often dubbed "emotional intelligence." But these jobs also pay lower salaries on the whole when set against jobs in industries like finance, banking, computer science, and engineering. Indeed, much care work isn't just low paid, it's frequently unpaid and invisible.[14] There are much-discussed sociological, cultural and historical reasons why women tend to cluster in such occupations, but the net result—combined with the fact that women also tend to be their children's primary caregivers, making them less likely to reach senior level positions than men—is the infamous gender pay gap.[15]

But could the stage be set for a reversal in women's fortunes? There is research to suggest that women may, in fact, fare a whole lot better than men from automation.[16] This is because it is generally agreed that jobs requiring literacy, social skills, and empathy are at less risk of automation than jobs requiring numeracy, calculation, and brute physical force. Even tasks like medical diagnosis and contract drafting—which involve the detection of patterns in information and mechanistic or process-driven problem-solving—are all far more vulnerable to automation and machine learning than tasks involving relationship building and "emotional and spiritual labor."[17]

Of course, it's one thing to say there might be more paid care work going around and more jobs available for women than men; it's quite another to say that care work will at long last be well paid. Wages are almost always determined by scarcity, and occasionally danger and difficulty. Empathy just isn't rare enough in our species or hard enough to do. True, an economic system that made compassion and well-being as fundamental to its perspective as want and scarcity would look very different. And perhaps more care work in a society would mark the difference between an aggressively competitive business culture and a more easy-going and cooperative culture. But there's no evidence that *just* by having more care work in an economy or holding it in higher esteem we'll stop paying people based roughly on how difficult it is to do what they do and on how many people are willing and able to do it. It's also just as likely that men will end up doing much more care work than has been typical thus far.[18]

The Nature of Work

Although studies modeling the *future* effects of AI on jobs and work have proved equivocal, studies looking at the *current* effects of such technologies are producing more concrete results. We know for example that job polarization is on the rise in many countries. Traditional middle-income jobs are giving way to high-value nonroutine cognitive work and low-value manual work.[19] Disconcertingly, there is some evidence suggesting that middle-income workers displaced by automation do not move into high-value nonroutine cognitive work.[20]

Another empirical approach is to look at incomes in areas in which new technologies have been deployed. We don't yet have evidence squarely focused on the effects of AI and machine learning, but we do when it comes to robots. Here the evidence suggests that the increasing use of robots has negative effects on both employment and wages.[21] These results are robust, as they seem to hold even when controlling for factors such as the availability of cheap imported goods, offshoring, and other influences affecting the decline in routine jobs.

A less well-known effect of the growing use of AI is its effect on working conditions. What little we do know doesn't paint a rosy picture. Consider those employed in the so-called gig economy. They are hired on very short-term contracts ("gigs") strictly on an as-needs basis via platforms such as Uber, Deliveroo, TaskRabbit, and Mechanical Turk. Although some of these workers are highly skilled and benefit from the flexibility of such work arrangements, a disproportionate number of them are poor, have little choice about the jobs they do, and little ability to negotiate better wages and conditions.[22] Many of them work in the tech sector itself (in case you thought all tech workers are well-paid twenty-somethings working for Apple, Facebook, and Google). They fill online orders in vast warehouses. They categorize Instagram status updates, videos, photos, and stories. They label images to help self-driving cars learn to detect cyclists and pedestrians. They photograph streets, detect homonyms, scan physical books, and classify violent and obscene social media posts. The repetitive and potentially dispiriting nature of much of this work belies the image many of us have of those "privileged" enough to work in the digital economy.

AI also sits behind many types of algorithmic management in which the day-to-day work of employees and contractors is assigned and assessed by algorithms. This is a disciplinary application of new technologies in which high-resolution monitoring and calibration of worker performance erodes what little agency, discretion, and autonomy a worker otherwise has.[23] Amazon has become the poster child for this regime. Its warehouses have been described as "the meatpacking assembly lines of our own age, where technological advances meld with capital's need to extract every last ounce of efficiency from its workforce."[24] Amazon employees have the speed of their work meticulously tracked, even as they are saddled with ridiculously unrealistic performance targets. In the United States, the company has been granted patents for "ultrasonic wristbands" that are sensitive to gesture and vibration (providing "haptic feedback") in an effort to monitor suboptimal performance.[25] The asphyxiating atmosphere created by this level of surveillance would make for dystopian entertainment were it not *actually* happening, or in the offing. Not that performance monitoring systems are necessarily reliable—they don't have to be in order to make someone's life miserable. As Adam Greenfield observes, "what is salient is not so much whether these tools actually perform as advertised, but whether users can be induced to believe that they do. The prejudicial findings of such 'HR analytics' ... may be acted upon even if the algorithm that produced them is garbage and the data little better than noise."[26]

As for the gig economy, even when workers receive positive ratings from their algorithmic bosses, these can effectively lock them into a single gig platform, as ratings aren't typically portable from one platform to the next.[27] On the other hand, at least in the gig economy, the flipside of controlling your workforce entirely by app—and hence at a physical distance—means that there is correspondingly less you can do to prevent organized resistance (as Uber and Deliveroo have found to their dismay in a growing number of countries).[28]

Why Work?

Let's suppose that the nature of work continues to evolve, with greater casualization, more "giggification," and higher levels of surveillance and control. It might be that these jobs represent the last *kind* of job in a mature developing digital economy, heralding a world in which there will be either much less work, very different work, or perhaps no real work at all to speak of. How should we feel about this? Before we rush to preserve current jobs or invent new ones, we'd do well to pause and consider what it is about work we value and what it is that all of us have a stake in salvaging from a world in which work provides the main means of subsistence. Obviously, we ought to welcome any technology that relieves us of onerous, dangerous, and demeaning work. But what about work that isn't onerous, dangerous, and demeaning? Could we do without work altogether? Is work in other ways beneficial? Is it virtuous or in any sense dignifying?

The idea that we could live well while working much less has been around for a long time, but it was most famously championed by the philosopher

Bertrand Russell in his 1932 essay, "In Praise of Idleness." Russell's claim that life would be better if most people worked about four hours a day is based on three ideas. First, that work is generally onerous. Given the option, most of us would do less of it. Second, that until relatively recently, most Europeans were subsistence farmers. The notion that work is virtuous and a source of dignity originates from an era in which virtually the sole imperative of humanity was to secure physical sustenance through personal toil—basically, tilling soil and growing food on common land. However, after the first industrial revolution, this mode of subsistence ceased to characterize the general lot of mankind, and innovative agricultural methods would see output reach levels not attainable through manual exertion alone. As Russell put it, the dignity of labor was an important idea before the industrial revolution but became an "empty falsehood" preached by the wealthy who "take care to remain undignified in this respect."[29] The final part of Russell's argument is that during the First World War, most of the English population was withdrawn from productive occupations and engaged one way or another in sustaining the war effort. Despite their engagement in this singularly wasteful and unproductive endeavor, he notes that the general level of physical well-being among unskilled wage earners on the side of the Allies was higher than before or since.

Russell concludes: "The war showed conclusively that, by the scientific organization of production, it is possible to keep modern populations in fair comfort on a small part of the working capacity of the modern world. If, at the end of the war ... the hours of work had been cut down to four, all would have been well. Instead of that the old chaos was restored."[30]

Farming and the production of the rest of what Russell calls the necessaries of life have, of course, become much more industrialized since 1932, and AI and robotics promise many more efficiencies in the near future. The reality of global warming is also calling into question the long-held assumption that the success of nations must be underpinned by the production of more and higher value goods and services, as reflected in higher GDP.[31] There is now actually an *inverse* relation between the wealth of OECD countries and the average number of hours worked by their citizens. The greatest average number of hours worked annually per citizen are in Mexico (2,250) South Korea (2,070), Greece (2,035), India (1,980), and Chile (1,970), whereas the fewest are in France (1,472), the Netherlands (1,430), Norway (1,424), Denmark (1,410), and Germany (,1363).[32]

Of course, as most economies are organized, Russell's prescription isn't one that could safely be followed by most of us today. But at a national level it does suggest that rationing work, perhaps by decreasing the length of the work week, might be beneficial in relatively well-off countries. But would a world with much less work be good for us? Despite the medieval origins of the notion that work is virtuous, is there really nothing virtuous, dignifying, or even edifying about work?

Probably part of the residual virtue that attaches to paid work attaches likewise to many forms of private income generation, even if these don't involve actual physical or mental labor (like passive income—rents, dividends, interest, and the like). The virtue here may derive from the belief that, by earning a living and making one's own way in life—however one manages to do this, be it through work or rent—one possesses the independence and maturity we associate with fully developed adulthood, at least in Western cultures. If this is true, though, the moral valuation involved obviously needn't be commensurate with any actual effort on the part of the earner.

Another part of what could make us think of work as virtuous may be that it frequently requires a degree of diligence, dedication, discipline, or at any rate structured living—qualities that may be perceived to be virtuous inasmuch as the opposites of these qualities indicate a lack of character, dependability, stability, or maturity. But such traits don't depend on the contract of employment for their existence. Anyone seriously invested in a personal project—poetry, printmaking, building, or renovating a home—needs to have a degree of diligence and discipline. Indeed, anyone out of work who's *looking* for work probably has to draw on greater reserves of such qualities than someone in regular, secure, long-term employment.

We could, if we wanted to, compile a list of the personal qualities any worker must be assumed to possess and chalk up the virtue or dignity of employment to any one of them. But we doubt the exercise will prove successful in isolating the *one* feature of paid work that *uniquely* explains its dignity or virtue. Truth is, it's not employment *per se* that's dignifying, virtuous, or edifying. If these labels can attach to anything, it's to those qualities we've already mentioned—diligence, dedication, etc.—*wherever* and *whenever* they're manifested, regardless of the existence of a formal contract of employment.

Still, there's no denying that for most people an employment contract will provide the usual forum in which such traits can be exercised, developed, and nurtured. One immensely valuable opportunity that formal employment affords fairly readily is social connection with colleagues. Although this varies from job to job and the connection isn't always positive, the opportunity for meaningful collaboration with others is terrifically important and under threat for many in the gig economy. Of course, one can find meaningful human connection outside work (obviously!), and, yes, one could join a community organization to experience the joys and sorrows of communal enterprise and shared identity. For that matter (and to repeat), one doesn't need formal work to pursue personal projects, form five-year plans, or establish long-term life goals. A life of leisure needn't conform to a stock stereotype, like hunting in the morning and fishing in the afternoon (as Karl Marx put it). But still, how you spend your time when the imperatives of survival have been met is really your answer to the question of what life is for. And that's a question that might call for at least a bit of personal reflection. Just like money, leisure must be thoughtfully spent. Russell decried the passive character of many popular amusements (attending the cinema, watching spectator sports, etc.).[33] He put it down partly to exhaustion—people are so tired from their daily exertions that they don't have the energy for anything vaguely strenuous at the end of a day's work—but also to a lack of proper training in how to use one's time wisely.[34] Taking a "discriminating pleasure in literature," for example, requires cultivation, practice, and perseverance. He's obviously not saying that spending our leisure time wisely is beyond us, by any means. But good quality leisure isn't a given, either, just because you happen to have the time for it. It takes "well-directed effort," as he put it.[35]

The bonds of employment allow many of us to defer these questions or at least not to have to consider afresh, day after day, what we want to do with our lives. In this sense, employment is valuable in the same way that having one local water company, postal service, or rail service, is valuable—it relieves us of the anxiety of choice. Not for a moment would we wish to romanticize, neutralize, or justify the evils of wage slavery. But a humane contract of employment (with hours compatible with the need for sleep, recreation, and family life, with generous benefits, paid holidays, and provisions for psychological and spiritual well-being) does at least provide many of us with the structures we'd otherwise have to put in place for ourselves, including the bonds of solidarity forged in the crucible of mutual endeavor.

10 Oversight and Regulation

What rules should govern artificial intelligence?

Unsurprisingly, it's a question that has attracted a wide range of answers. In recent years, however, a rough consensus seems to be emerging that there *should* be rules. The consensus doesn't just include the usual suspects, either—academics, activists, and politicians. In 2014, tech entrepreneur and celebrity Elon Musk surprised many by calling for "some regulatory oversight" of AI, "just to make sure that we don't do something very foolish."[1] Facebook CEO Mark Zuckerberg's calls are similar (though made for different reasons).[2]

But what form should this regulatory oversight take? What should be its target? Who should be doing the overseeing? And what sorts of values or standards should they be seeking to enforce?

It's hardly an original observation that emerging technologies present challenges for lawmakers and regulators. When we don't know for sure what form a technology will take, how it will be used, or what risks it will pose, it can be hard to tell whether existing rules will be adequate or whether a new regulatory scheme will be required instead. The challenge is especially daunting when the technology in question is something like AI, which is actually more like a family of diverse but related technologies with an almost unlimited range of uses in an almost unlimited range of situations. What sorts of rules or laws, we might wonder, are supposed to manage something like *that*? Rules, after all, are—pretty much by definition—relatively certain and predictable. A rule that tried to keep pace with a fast-evolving technology by constantly changing? That, you might think, would hardly qualify as a rule at all.

Luckily, AI is by no means the first emerging technology to have posed these challenges. Decisions about the right rules can be made against a background of considerable experience—good and bad—in responding to predecessor technologies: information and communication technologies, for example, as well as genetic and reproductive technologies.

What Do We Mean by "Rules"?

So far, we have used the terms "rules," "regulation," and "laws" interchangeably and fairly loosely. But what exactly are we talking about? There's actually quite a wide range of options available to regulate AI. On one end of the spectrum, there are relatively "hard-edged" regulatory options. Predominantly, we're talking about regulation by *legal* rules.

When most people think about "legal rules," they probably think in terms of legislation, "the law on the statute books." They might also consider rulings by courts. Both of these are important sources of legal rules (the *most* important, really). But legal rules can come in a wider variety of forms. They can, for instance, be made by regulatory agencies with delegated responsibility. In the realm of assisted reproductive technologies, organizations like the UK's Human Fertilization and Embryology Authority and New Zealand's Advisory Committee on Assisted Reproductive Technology (ACART) are empowered by statute to set limits and conditions on their use, at least up to a point.

"Regulation" is an even wider concept than that. It's a much discussed and somewhat contested term in the academy, but it's still often thought to be a wider concept than just "law." For the purposes of this book, we've taken a fairly open-ended approach to what counts as "regulation." The questions raised by the emergence of AI are so many and so diverse that it would be rash to rule out any possible answers before we even start. For our purposes then, "regulation" means laws and court judgments, but it doesn't just mean that. Like AI itself, the rules we apply to it could come in a wide and varied array of forms.

Who Makes the Rules?

Regulation then, needn't be confined to legal prohibitions and orders, and neither is it limited to rules issued by legislatures, courts, or regulatory agencies. What other forms could it take? A commonly proposed alternative in the context of AI is that of *self-regulation*. Companies might set rules for themselves, or industry bodies might elect to impose some rules upon their members.

Why would a company or a whole industry opt to do this, to tie its own hands in such a way? Cynically, we might see it as a self-interested alternative to having it imposed from outside. In other cases, other motives might play a part. It might be to offer reassurance to clients and customers, for example, or to establish and uphold the reputation of that industry as a safe and responsible one. Some critics of the adequacy of self-regulation are more overtly skeptical of corporate motivations. Cathy Cobey, writing in *Forbes* magazine, suggests that even when companies call for externally imposed regulations, the real reason stems from a concern that, "if the decisions on how to govern AI are left to them, the public or the court system may push back in the years to come, leaving corporations with an unknown future liability."[3]

To what extent can self-regulation adequately protect society from the worst risks of AI? For many commentators, this just doesn't offer enough by way of guarantees. James Arvanitakis has argued that "for tech companies, there may be a trade-off between treating data ethically and how much money they can make from that data." This has led him to conclude that, "in order to protect the public, external guidance and oversight is needed."[4] Jacob Turner has made a similar point, explaining that "considerations of doing good are often secondary to or at the very least in tension with the requirement to create value for shareholders."[5]

For Turner, another weakness with self-imposed rules is that they lack the binding nature of actual law. "If ethical standards are only voluntary," he suggests, "companies may decide which rules to obey, giving some organizations advantages over others."[6]

Recently, even some major industry players seem to be coming round to the view that, although self-regulation will have a major role to play, there is also a role for governments, legislatures, and "civil society" in setting the rules. In its 2019 "white paper," Google combined a fairly predictable preference for self- and co-regulatory approaches, which it claimed would be "the most effective practical way to address and prevent AI related problems in the vast majority of instances," with an acknowledgment that "there are some instances where additional rules would be of benefit."[7]

Interestingly, Google's position was that this benefit was "not because companies can't be trusted to be impartial and responsible but because to delegate such decisions to company uses would be undemocratic."[8] This relates to what some emerging technology commentators refer to as "regulatory legitimacy." This has been explained as meaning that "as technology matures, regulators need to maintain a regulatory position that comports with a community's views of its acceptable use."[9] Simply leaving the rules up to industry to determine is unlikely to satisfy this demand, at least in areas where the technology is likely to have important impacts at an individual or societal level.

Google's preferred solution seems to involve a "mixed model," combining self-regulation with externally imposed legal rules. Precedents aplenty exist for such a hybrid. In many countries, the news media is governed by self-regulatory mechanisms, such as the UK's Press Complaints Commission, Australia's Press Council, or New Zealand's Media Council. But news media are also subject to "the law of the land" when it comes to matters like defamation law, privacy, and official secrets. The medical profession is another example of a mixed model, subject to "the law of the land" in many respects, but also to professional ethics and internal disciplinary structures. Doctors who break the rules of the profession can be suspended, restricted, or even "struck off," but doctors who break the laws of society can also be sued or occasionally even imprisoned.

Whether we think that would be a desirable model for AI might depend on how well we think it has functioned in other industries. Most suspicion probably arises when self-regulation functions as an *alternative* to legal regulation rather than as an adjunct to it. And there's obviously nothing to stop a company or a whole industry opting to set rules that go *further* than the existing law.

Flexibility

Regulation can also come in both more and less flexible forms. Very specific rules will allow little room for interpretation. To go back to the area of reproductive technologies, both UK and New Zealand laws contain a number of flat-out bans on certain practices—creating animal-human hybrids for instance or human cloning.

In the context of AI, there have been several calls for outright bans on certain kinds of applications. For example, in July 2018, an open letter was published by the Future of Life Institute, addressing the possibility of "[a]utonomous weapons [that] select and engage targets without human intervention." The open letter, which has now been signed by over 30,000 people, including 4,500 AI and robotics researchers, concluded that "starting a military AI arms race is a bad idea and should be prevented by a ban on offensive autonomous weapons beyond meaningful human control."[10]

Other specific applications of AI technology have also been the subject of actual or suggested bans. Sometimes these are context-specific. For example, California has enacted a (temporary) ban on facial recognition on police body cams,[11] whereas San Francisco has gone further by banning the technology's use by all city agencies.[12]

These are examples of *negative* obligations, rules that tell those who are subject to them that they must avoid acting in certain ways. Rules can also impose *positive* obligations. California again provides a good example. The Bolstering Online Transparency Bill (conveniently abbreviating to "BOT"), which came into effect in July 2019, "requires all bots that attempt to influence California residents' voting or purchasing behaviors to conspicuously declare themselves."[13]

How effective any of these new laws will prove to be at achieving their objectives remains to be seen, but those objectives are quite clear and specific. Regulation needn't always be quite so clear-cut. It can also take the form of higher level guidelines or principles. These are still important, but they come with a degree of flexibility in that they have to be interpreted in light of particular circumstances.

There have been many examples of this sort of "soft law" for AI. The EU Commission's recent AI principles are of this kind, but we've also seen guidelines issued by the OECD,[14] the Beijing Academy of Artificial Intelligence,[15] the UK's House of Lords Select Committee on AI,[16] the Japanese Society for Artificial Intelligence,[17] and the Asilomar Principles from the Future of Life Institute.[18] Google and Microsoft have also issued statements of principle.[19] One thing we can safely say is that, if anything goes badly wrong with AI, it probably won't be for want of guidelines or principles!

Examining these various documents and declarations reveals a fair degree of agreement. Almost all of them identify principles like fairness and transparency as being important, and most insist that AI be used to promote something like human well-being or flourishing. This might seem like a promising sign for international consensus. On the other hand, we could say that the principles and the agreement are at a very high level of generality. Would anyone *disagree* that AI should be used for human good or argue that it should be unfair? For lawyers and regulators, the devil will be in the detail, in turning these commendable but very general aspirations into concrete rules and applying them to real life decisions. They'll also want to know what should happen when those principles conflict with one another, when transparency clashes with privacy maybe, or fairness with safety.

Sometimes the firmer and more flexible rules work together. As well as clear and specific bans on things like human cloning, New Zealand's Human Assisted Reproductive Technology Act contains general guiding principles and purposes that the regulators must keep in mind when making decisions. These include things like "the human health, safety, and dignity of present and future generations should be preserved and promoted" and "the different ethical, spiritual, and cultural perspectives in society should be considered and treated with respect." These principles must inform how the regulators act, but they are open to a range of interpretations in the context of particular decisions.

Rules can be binding. That's probably true of most of the rules that come readily to mind. The laws against drunk driving, burglary, and tax fraud aren't discretionary. But rules can also be advisory, perhaps setting out "best practice" standards. The binding vs. nonbinding classification is different from the specific vs. generic one. A rule can be specific (say, about how to use a particular type of software), but merely advisory. Programmers working on that specific software could choose to ignore the recommendation without penalty. On the other hand, rules could be quite general or vague (like the principles we just mentioned) and yet mandatory. Programmers would *have* to follow them (even if there's wiggle room on interpretation).

We can see then that "regulation" is a fairly wide concept. It can come from a range of sources—from governments, legislatures, and courts, certainly, but also from agencies delegated to regulate particular areas and even from industries or companies themselves. Its form can range from the very specific (bans on particular uses or practices) to more high-level principles. Finally, its effect can range from the binding, with legal penalties for noncompliance, to the advisory and merely voluntary.

Any of these approaches—or a combination of them—could be used to regulate AI. The next question concerns what exactly it is we're trying to regulate in the first place.

The Definition Problem: What Are We Even Talking About?

The first task for anyone proposing regulation probably involves defining what exactly it is they want to regulate. That's no big challenge when the target is something like "Lime scooters" or "laser pointers," but when the target is something as broad as "AI," the task can be daunting.

This absence of a "widely accepted definition of artificial intelligence" has been a recurrent theme in the literature,[20] and at first glance, it does look like a major stumbling block. Courts, after all, need to know how to apply new rules, regulators need to know the boundaries of their remit, and stakeholders need to be able to predict how and to what the rules will apply.

Is it actually necessary to have a definition of "AI" before we can begin to assess the need for regulation? That might depend on a couple of factors. For one thing, the need for a precise definition is likely to depend a lot on whether we're attempting to create AI-specific rules or regulatory structures. If laws of more general application would be adequate in this context, then the definitional challenge becomes less pressing because whether or not something qualifies as "AI" will have little bearing on its legal status. But if we think we need AI-specific laws and regulations, then we certainly would need a definition of what they apply to.

Another possible response to the definitional challenge is to forswear an all-purpose definition and adopt a definition suited to a particular risk or problem. Rules could be introduced for predictive algorithms, facial recognition, or driverless cars without having to worry unduly about a general definition of "AI."

Regulatory Phase: Upstream or Downstream?

If we can agree on *what* and *how* they want to regulate, regulators then have to decide on the question of *when*. In some cases, the advantages of very early regulation—before the technology has been brought to market or indeed even *exists*—can be obvious. If a technology is seen as particularly dangerous or morally unacceptable, it may make sense to signal as much while it's still at a hypothetical stage. In less extreme cases, early intervention might also be beneficial, shaping the direction of research and investment. Gregory Mandel has argued that intervention at an early stage might meet less stakeholder resistance, as there will predictably be fewer vested interests and sunk costs, "and industry and the public are less wed to a status quo."[21]

Alternatively, there may be times when it's better to wait and see, dealing with problems if and when they arise rather than trying to anticipate them. The internet is often held up as an example of a technology that benefitted from such an approach. Cass Sunstein has made the case for caution before rushing to regulate, pointing out that, at an early, "upstream" stage, it's harder to make accurate predictions and projections about costs and benefits. "If we will be able to do so more accurately later on, then there is a (bounded) value to putting the decision off to a later date."[22]

There's almost no straightforward answer to whether it's better to regulate early or late. Everything depends on the particulars of the technology. Luckily, we are not faced with a binary choice. The decision is not "everything now" or "nothing until later." It's entirely possible to take regulatory steps to address some risks early—maybe those that are particularly serious, immediate, or obvious—and to defer decisions about those with are more speculative or distant until later when we are better informed.

Regulatory Tilt and Erring on the Safe Side

In many cases, however, there is just no way to defer all decision-making until later. When a technology exists in the here and now and is either being used or the subject of an application for use, the regulatory option of putting off the decision just isn't on the table.

What sort of approach should regulators adopt when making decisions in the face of uncertainty? The idea of "regulatory tilt" describes the starting point or default setting that they should adopt. If they can't be sure of getting it right (and they will often be in doubt), in which direction should they err?

One obvious approach to this question is that regulators faced with uncertainty should err on the side of caution or safety. This is sometimes expressed in terms of the precautionary principle. When the EU Parliament made its recommendations on robotics to the EU Commission in 2017, it proposed that research activities should be conducted in accordance with the precautionary principle, anticipating potential safety impacts and taking due precautions.[23]

When an emerging technology presents an uncertain risk profile, there does seem to be something appealing about the application of a precautionary approach. Waiting for conclusive evidence about the dangers it presents may result in many of those risks materializing, causing preventable harm or loss. In some cases, the loss may be of a nature that can't easily be put right, not only in terms of individual people harmed, but in the sense that it may be impossible to put the genie back in the bottle. Think of a genetically modified bacterium or runaway nanobot released into the wild. Some of the concerns about runaway superintelligent AI are of that nature. If the concerns are at all credible, then they are of the stable door variety; there's not much to be done once the proverbial horse has bolted.

Nick Bostrom is probably the best-known voice of caution about such risks. "Our approach to existential risks," he warns, "cannot be one of trial-and-error. There is no opportunity to learn from errors."[24] Bostrom has proposed what he calls a "maxipok" approach to such risks, which means that we act so as to "maximize the probability of an okay outcome, where an 'okay outcome' is any outcome that avoids existential disaster."[25]

When the risks are of an existential nature—as some commentators really seem to believe they are where AI is concerned—the case for precaution looks pretty compelling. The challenge, of course, is to determine when such risks have any credible basis in reality. Our legal and regulatory responses, presumably, should not be responses to the direst dystopian visions of futurists and science fiction writers. But how are regulators meant to tell the real from the fanciful? History—indeed the present!—is littered with horror stories of major risks that were overlooked, ignored, or disguised, from asbestos to thalidomide to anthropogenic climate change and global pandemics. Less well known, perhaps, is that history is also replete with examples of worries about new technologies that amounted to nothing—that to modern eyes, look quite ridiculous. Cultural anthropologist Genevieve Bell has recounted that, in the early days of railway, critics feared "that women's bodies were not designed to go at fifty miles an hour," so "[female passengers'] uteruses would fly out of [their] bodies as they were accelerated to that speed"![26]

Some cases are genuinely very hard for regulators because the science involved is both demanding and contested. Shortly before the Large Hadron Collider (LHC) was turned on at CERN, an attempt was made by some very concerned scientists to obtain a court order preventing its initiation.[27] The basis of their claim was a concern that the LHC could create a black hole, posing an existential threat to the planet. It is hard to envy the judge, trying to weigh up competing claims from theoretical physicists, most of whom thought the operation safe, but a minority of whom believed it could spell the end the world.

In the event, the court opted to trust the majority view, the supercollider was switched on, and (as far as we know) nothing catastrophic occurred. However, an approach that prioritized the minimization of existential risk above all else would presumably have erred in the other direction.

Even where we have some good reason to take seriously a major warning, it won't always be the case that the "moral math" produces an obvious answer about how to respond. Often, decisions to avoid some risks will mean foregoing certain kinds of benefits. In the case of some forms of AI, those benefits could be very considerable. A precautionary approach to driverless cars, for instance, would certainly eliminate the risk of deaths by driverless cars. But an estimated 1.2 million people currently die on the roads every year. If driverless cars had the potential to reduce that number quite dramatically, passing up that chance would amount to quite a risk in itself.

It's easy to think of other potential applications of AI that might present the same kind of trade-off. For example, in medical diagnostics the possibility exists that AI could do considerably better than humans and that better performance could be translated into lives prolonged or improved.

What about the sort of superintelligent AI that seems to worry Bostrom and Musk, though? Maybe there's a case for a precautionary ban even on researching that sort of thing. But again the moral math dictates that we consider what we may be passing up in so doing. What avenues of research would need to be closed off in order to prevent someone even accidentally taking us into the realms of superintelligence? Might that mean that we had to pass up potential AI solutions for hitherto intractable problems, maybe even something as serious as climate change?[28] Is the threat of a runaway superintelligence really more existentially pressing than that?

A simple rule that says "always avoid the worst outcome" might seem appealing at first glance, but it isn't clear how to err on the side of safety when there may be existential threats in all directions or when the alternative seems less bad but more likely. Looked at like that, it seems that there really is no simple heuristic that will guarantee we avoid the worst scenario.

An FDA for AI?

Laws enacted by legislatures are certainly an important source of law. They offer certain advantages. They have democratic legitimacy, being enacted by the elected representatives of that society. They are likely to be highly transparent (it's hard to pass a law without anyone noticing), which means that, ideally, their effects should be predictable.

But on the deficit side, legislation is notoriously slow to adapt. Getting a new law through the legislature can take a long time.* And since trying to anticipate every possible situation in a few pages is notoriously difficult, much of the new law's effect will only be decided when courts come to apply it in particular situations.

An alternative—or supplementary—option could be a specialized regulatory agency. Matthew Scherer has argued that regulatory agencies have several advantages when dealing with an emerging technology. For one thing, agencies "can be tailor-made for the regulation of a specific industry or for the resolution of a particular social problem."[29] Although legislatures will be populated by generalists, members of agencies can be appointed on the basis of their specialist knowledge in a particular area. For example, New Zealand's assisted reproduction regulator, ACART, is required by statute to have at least one member with expertise in assisted reproductive procedures and at least one expert in human reproductive research.

Regulatory agencies, Scherer argues, also have an advantage over court-made rules. Judges are significantly limited in their remit, being restricted to making decisions about the cases that actually appear before them. Regulatory agencies aren't similarly constrained.

Regulatory agencies come in a wide variety of forms and have a wide array of remits and responsibilities. As with legislation, they could be specific to a particular technology or family of technologies—ACART and the UK's HFEA, for instance, are focused on reproductive technologies. Or they could be fashioned with a particular policy objective or value in mind—the UK's Information Commissioner's Office and New Zealand's Office of the Privacy Commissioner were established to focus specifically on privacy and data protection.

---

*Often, but not always. On March 15, 2019, a mass shooting in the New Zealand city of Christchurch was live-streamed on Facebook. The Australian Government introduced legislation that required social media platforms to "ensure the expeditious removal" of "abhorrent violent material" from their content service. The Criminal Code Amendment (Sharing of Abhorrent Violent Material) Act 2019 came into effect on April 6, a mere three weeks after the event to which it was a response. The Government has been criticized for its failure to consult on the content or likely effects of the Act, a regular criticism of particularly fast-track lawmaking.

Regulatory agencies can also be conferred a wide array of powers. Some have compulsory inspectorate functions; others are able to hand out penalties and sanctions. Some can create rules, whereas others exist to enforce or monitor compliance with rules created by others. Often, they will operate at a "softer" level, for example, in issuing best practice guidelines or codes of practice or simply giving advice when requested. Which of these models would be best suited for the context of AI is obviously going to be an important consideration.

An example of an existing regulatory setup that has recently attracted some interest as a possible model for AI and algorithms is in the pharmaceutical industry. Although specifics vary between jurisdictions, most countries have some sort of regulatory agency in place. Perhaps the best known is the United States' Food and Drug Administration (FDA). Several writers have proposed the idea of a sort of FDA for AI.[30]

As Olaf Groth and his colleagues note, not every application of AI technology would need to be scrutinized by the regulatory agency. Instead, the agency "would need distinct trigger points on when to review and at what level of scrutiny, similar to the ways the FDA's powers stretch or recede for pharmaceuticals versus nutritional supplements."[31]

But Andrea Coravos and her colleagues are skeptical of the notion of a single AI regulator that could operate across all disciplines and use cases. Instead, they suggest that "oversight can and should be tailored to each field of application."[32] The field of healthcare, they claim, would be one field "already well positioned to regulate the algorithms within its field," whereas "other industries with regulatory bodies, such as education and finance, could also be responsible for articulating best practices through guidance or even formal regulation."[33] Coravos and her colleagues may have a point, but it's not clear how well their modular institutional setup could handle some of the novel conundrums thrown up by a radically innovative tech sector. We have talked, for example, about the risk that "bots" or recommender algorithms in social media could influence the outcomes of elections, or potentially lead people to increasingly extreme political positions. These aren't areas of life that have so far been subject to much in the way of regulatory scrutiny.*

*There are, of course, rules about influencing politics, but these are typically about matters like donations to campaigns or candidates—and even those are fairly minimal in some jurisdictions (Citizens United). But individually targeted campaign

In 2019, researchers at New Zealand's Otago University (including some of the authors of this book) proposed the creation of a specialist regulator to address the use of predictive algorithms in government agencies.[34] How would such an agency function? In an article that also made the case for an FDA-based model, Andrew Tutt considered a range of possible functions. These occupy a range of places on a scale from "light-touch" to "hard-edged." The agency could, for example

- act as a standards-setting body,
- require that technical details be disclosed in the name of public safety, and
- require that some receive approval from the agency before deployment.[35]

The last of these suggestions would, on Tutt's analysis, be restricted for the most "opaque, complex, and dangerous" uses. It could "provide an opportunity for the agency to require that companies substantiate the safety performance of their algorithms." Tutt also suggests that premarket approval could be subject to usage restrictions; as with drugs, distribution without permission or "off-label" use of AIs could attract legal sanctions.

Pre-deployment approval is worth taking seriously. Regulation pitched at the level of particular AI algorithms or programs seems likely to overlook the fact that these are in many cases highly flexible tools. An AI approved for an innocuous use could be repurposed for a much more sensitive or dangerous one.

Rules *for* AI?

The rules we have discussed so far are rules *about* AI, but they are rules that will be applied to the human beings that manufacture, sell, or use AI products. What about rules programmed *into* AIs? Is there a case for requiring certain things to be required or prohibited?

Again, most of the attention to date has focused on embodied AIs, like driverless cars and robots of various sorts, for which the risks of harm are most obvious. In the context of driverless cars, much attention was garnered by the apparent acknowledgment by Mercedes-Benz that they would program their cars to prioritize the lives of the occupants over any other

---

adverts and chatbots purporting to be human contributors to online discussions have thus far avoided much in the way of regulatory scrutiny—San Francisco's new BOT law being a very rare example.

at-risk parties.[36] The logic, expressed in an interview by Christoph von Hugo, manager of driver assistance systems and active safety, was ostensibly based on probability. "If you know you can save at least one person, at least save that one," he said. "Save the one in the car."[37]

There may be something initially attractive about this bird-in-the-hand logic, but the moral math may not so readily support it. What if the cost of saving "the one in the car" is the likely deaths of the ten on the pavement? Or the thirty on the school bus? How much more confident would the AI have to be of saving the occupant for such a judgment to be justified?

More cynically, we may wonder how many driverless cars Mercedes or any other manufacturer would sell if they promised anything else. A study published in *Science* a few months earlier had demonstrated that, although a significant number of people recognized the morality of sacrificing one occupant for ten pedestrians, far fewer wanted this for their own car.[38]

You may not find this outcome entirely surprising. Prioritizing our own well-being and that of those close to us is a common trait with obvious evolutionary benefits. The authors certainly were not surprised. "This is the classic signature of a social dilemma," they wrote, "in which everyone has a temptation to free-ride instead of adopting the behavior that would lead to the best global outcome."[39] A typical solution, they suggested, "is for regulators to enforce the behavior leading to the best global outcome." Should we have rules to prevent free riding in cases like this?

The idea of sharing the road with vehicles programmed to prioritize the lives of their own customers over everyone else seems to strike at the core of several of our moral values—an egalitarian concern for the equal value of lives, as well as a utilitarian concern with minimizing harm. It may also violate strong ethical (and legal) rules against active endangerment. The car that swerves to hit a cyclist or a pedestrian to save its occupant wouldn't just be failing to make a noble sacrifice; it would be actively endangering someone who may not otherwise have been in danger. Although criminal law has typically been slow to blame human drivers who flinch from altruistic self-sacrifice in emergencies, it may well take a different view of someone who makes such choices in the calm surroundings of a computer lab or a car showroom.

Other survival priorities, though less obviously self-serving, could be equally challenging to what we might consider our shared values. A 2018 ("moral machine") study found an intriguing range of regional and cultural differences to driverless vehicle dilemmas across a range of countries.[40]

Some common preferences (for example, sparing young people over old) are intelligible, if controversial. Some (the preference for prioritizing female lives over male) might seem archaic. And some (sparing those of higher over lower social status) are likely to strike many people as simply obnoxious.

These studies, though somewhat artificial, might suggest a need for rules to ensure that, when choices have to be made, they are not made in ways that reflect dubious preferences. Indeed, Germany has already taken steps in this direction. In 2017, the Ethics Commission of its Federal Ministry of Transport and Infrastructure published a report on automated and connected driving. The report specifically addressed dilemma cases and the sorts of rules that could be programmed for such eventualities—"In the event of unavoidable accident situations, any distinction based on personal features (age, gender, physical, or mental constitution) is strictly prohibited."

The report also said that, although "[g]eneral programming to reduce the number of personal injuries may be justifiable," it would not be permissible to sacrifice parties not "involved in the generation of mobility risks" over those who are so involved.[41] In other words, innocent pedestrians, bystanders, cyclists, and so on are not to be sacrificed to save the occupants of autonomous vehicles.

Autonomous vehicles might be the most obvious current example of this kind of challenge, but they are unlikely to be the last. Decisions that until now have been too rare or too instinctive to merit much by way of a legal response are now going to be made in circumstances in which rules can meaningfully be brought to bear. But as the "moral machine" study shows, regulators might struggle to find an ethical consensus about what rules we should agree upon, and the research to date suggests that leaving these decisions to market forces won't guarantee an outcome that's satisfactory to most of us.

Summing Up

We need new rules for AI, but they won't always need to be AI-specific. As AI enters almost every area of our lives, it will come into contact with the more general rules that apply to commerce, transport, employment, healthcare, and everything else.

Nor will we always need *new* rules. Some existing laws and regulations will apply pretty straightforwardly to AI (though they might need a bit of tweaking here and there). Before we rush to scratch-build a new regulatory regime, we need to take stock of what we already have.

AI will, of course, necessitate new AI-specific rules. Some of the gaps in our existing laws will be easy to fill. Others will force us to revisit the values and assumptions behind those laws. A key challenge in fashioning any new regulatory response to technology is to ensure that it serves the needs of all members of society, not just those of tech entrepreneurs and their clients, or indeed those of any other influential cohort of society.

Epilogue

We started writing this book in 2019. Many people, including some governments, were becoming aware of the challenges posed by AI technologies, and the movement to regulate to minimize their risks was gaining momentum. The prospects of being subject to algorithmic decisions that are hard to challenge or even understand, of losing your job to an AI, or of being driven by driverless cars were, and still are, matters of great interest and concern.

But we started writing this book when none of us had heard the term "COVID-19."

As we put the finishing touches on this book, we look out on a world still in the midst of a pandemic. The numbers infected are approaching one and a half million, with 75,000 deaths. Entire countries are locked down, with an estimated one-third of the world's population under some kind of restriction, many largely confined to their homes. Emergency laws are being rushed into effect and desperate measures being taken to plug gaps in health care provision and forestall economic collapse.

It's a fool's errand to try to predict the nature and scale of the changes that COVID-19 will leave behind. It seems safe to say, though, that for many people, societies, and governments, the epidemic of 2020 will lead to a significant re-evaluation of priorities.

These changed priorities will be reflected in regulatory responses. In late March 2020, the *Financial Times* quoted an anonymous source "with direct knowledge of the European Commission's thinking" as saying that "the EU is not backtracking yet on its position but it is thinking more actively about the unintended consequences of what they have proposed in the white paper on AI."[1]

We can only speculate about what unintended consequences they have in mind, but it's easy to imagine that the balance between safety and privacy may be struck differently in the context of the present crisis. Whether AIs can do jobs just as well as humans may be seen as less important than whether they can offer some sort of help when there are no humans available. And rigorous testing and scrutiny of algorithmic tools may seem like a luxury when they promise the chance of early detection or contact tracing of infected people, triaging scarce resources, perhaps even identifying treatments.

It's understandable—inevitable, really—that in the immediate grip of this crisis, attention is focused on whatever can slow the disease's spread and provide essential treatment for those in need. But as we emerge from months of lockdown into potentially far longer periods of restriction, surveillance, and rationing, hard questions about privacy, safety, equity and dignity will have to be asked. As governments and police assume new powers, and technology is rapidly pressed into service to track and monitor our movements and contacts, questions about accountability and democracy can't be ignored. As we noted in the prologue, the technologies currently being used to contact-trace infected people—and, perhaps more controversially, to police quarantine compliance—won't disappear once the crisis has passed. It is naïve to expect governments, police forces, and private companies to hand those powers back.

The future of work, too, might look very different. The COVID-19 crisis has taught us how vital many precarious workers—delivery drivers, cleaners, shelf-stackers—are to the functioning of our societies. Yet as companies scramble for survival in what seems very likely to be a prolonged recession, it is those very workers whose livelihoods could be endangered if technological replacement seems more economically viable. As we write, Spain has announced its intention to introduce a permanent universal basic income, a measure that only months ago was widely viewed as experimental and likely unaffordable.[2] All around us, the scarcely imaginable is fast becoming the seriously entertained.

Decisions about how we respond to these challenges will need to be informed by technical, ethical, legal, economic and other expertise. The populist refrain that "we've heard enough from experts" is surely going to fade, at least for a while. But these sorts of decisions can't be made just by experts. Technocracy is not democracy. Neither is an oligarchy of wealthy tech entrepreneurs.

If we are to leverage the benefits of AI technologies while side-stepping the pitfalls, it's going to take vigilance. Not just vigilance from governments and regulators, though we firmly believe that this will be necessary. And not just vigilance from activists and academics, though we hope that we've made a modest contribution in this book. But vigilance from all sectors of society: from those who are going to be on the sharp end of algorithmic decisions and those whose jobs are going to be changed or replaced altogether.

It's a tall order for people who are already struggling with financial insecurity, discrimination, state oppression, or exploitative employer practices to take the time to learn about something like artificial intelligence. The field is moving at a discombobulating pace. But for those citizens who do have the resources and inclination, we wish you well. Your role as citizen scrutinizers will be essential to keeping the technology and its applications fair. We hope this book will support your efforts.

About the Authors

John Zerilli is a philosopher with particular interests in cognitive science, artificial intelligence, and the law. He is currently a Research Fellow at the Leverhulme Centre for the Future of Intelligence at the University of Cambridge, and from 2021 will be a Leverhulme Trust Fellow at the University of Oxford.

John Danaher is Senior Lecturer in the School of Law, National University of Ireland, Galway.

James Maclaurin is Professor of Philosophy at the University of Otago, New Zealand.

Colin Gavaghan is Professor of Law at the University of Otago, New Zealand.

Alistair Knott is Associate Professor in the Department of Computer Science, University of Otago, New Zealand.

Joy Liddicoat is a lawyer specializing in human rights and technology, and a former Assistant Commissioner at the Office of the Privacy Commissioner, New Zealand.

Merel Noorman is Assistant Professor in AI, Robotics, and STS at the University of Tilburg, the Netherlands.

Notes

Prologue

1. A. M. Lowe, "Churchill and Science," in *Churchill by His Contemporaries*, ed. Charles Eade (London: Reprint Society, 1955), 306.

2. Jamie Susskind, *Future Politics: Living Together in a World Transformed by Tech* (New York: Oxford University Press, 2018), 54.

3. Richard Susskind and Daniel Susskind, *The Future of the Professions: How Technology Will Transform the Work of Human Experts* (Oxford: Oxford University Press, 2015), 50.

4. Tim Hughes, "Prediction and Social Investment," in *Social Investment: A New Zealand Policy Experiment*, ed. Jonathan Boston and Derek Gill (Wellington: Bridget Williams Books, 2018), 162.

5. Paul E. Meehl, *Clinical Versus Statistical Prediction: A Theoretical Analysis and a Review of the Evidence* (Minneapolis: University of Minnesota Press, 1954); R. M. Dawes, D. Faust, and P. E. Meehl, "Clinical Versus Actuarial Judgment," *Science* 243, no. 4899 (March 1989): 1668–1674; W. M. Grove et al., "Clinical Versus Mechanical Prediction: A Meta-Analysis," *Psychological Assessment* 12, no. 1 (April 2000): 19–30; J. Kleinberg et al., *Human Decisions and Machine Predictions* (Cambridge, MA: National Bureau of Economic Research, 2017).

6. Hughes, "Prediction and Social Investment," 164–165.

7. M. Ribeiro, S. Singh and C. Guestrin, "'Why Should I Trust You?' Explaining the Predictions of Any Classifier," *Proc. 22nd ACM International Conference on Knowledge Discovery and Data Mining* (2016): 1135–1144.

8. Steve Lohr, "Facial Recognition Is Accurate, if You're a White Guy," *New York Times*, February 9, 2018.

9. Stuart Russell and Peter Norvig, *Artificial Intelligence: A Modern Approach*, 3rd ed. (Upper Saddle River, NJ: Prentice Hall, 2010), 3.

10. Luciano Floridi, *The Fourth Revolution: How the Infosphere is Reshaping Human Reality* (Oxford: Oxford University Press, 2014).

11. *Associated Provincial Picture Houses Ltd. v. Wednesbury Corporation*, [1948] 1 K.B. 223.

12. Alvin Toffler, *Future Shock*, British Commonwealth ed. (London: Pan Books, 197), 399.

13. Ibid., 398.

Chapter 1

1. Marvin Minsky, ed. *Semantic Information Processing* (Cambridge, MA: MIT Press, 1968).

2. Nigel Watson, Barbara Jones, and Louise Bloomfield, *Lloyd's Register: 250 Years of Service* (London: Lloyd's Register, 2010).

3. Maurice Ogborn, *Equitable Assurances: The Story of Life Assurance in the Experience of the Equitable Life Assurance Society 1762–1962* (London: Routledge, 1962).

4. Samuel Kotz, "Reflections on Early History of Official Statistics and a Modest Proposal for Global Coordination," *Journal of Official Statistics* 21, no. 2 (2005): 139–144.

5. Martin Clarke, "Why Actuaries Are Essential to Government," *UK Civil Service Blog*, December 4, 2018, https://civilservice.blog.gov.uk/2018/12/04/why-actuaries-are-essential-to-government.

6. Martin H. Weik, "The ENIAC Story," *Ordnance, the Journal of the American Ordnance Association* (January–February 1961): 3–7.

7. George W. Platzman, *The ENIAC Computation of 1950: Gateway to Numerical Weather Prediction* (Chicago: University of Chicago Press, 1979).

8. Judy D. Roomsburg, "Biographical Data as Predictors of Success in Military Aviation Training," Paper presented to the Faculty of the Graduate School of the University of Texas at Austin, December 1988.

9. Wei-Yin Loh, "Fifty Years of Classification and Regression Trees," *International Statistical Review* 82, no. 3 (2014): 329–348.

10. Frank Rosenblatt, *The Perceptron: A Perceiving and Recognizing Automaton* (Buffalo: Cornell Aeronautical Laboratory, 1958).

11. Donald Hebb, *The Organization of Behavior: A Neuropsychological Theory* (Oxford: Wiley, 1949).

12. Alex Krizhevsky et al., "ImageNet Classification with Deep Convolutional Neural Networks," *Communications of the ACM* 60, no. 6 (May 2017): 84–90.

Chapter 2

1. Parts of this chapter are reprinted with permission from Springer Nature, *Philosophy and Technology*, "Transparency in Algorithmic and Human Decision-Making: Is There a Double Standard?" by John Zerilli, Alistair Knott, James Maclaurin, and Colin Gavaghan. Copyright 2018. As such, it represents the considered views of these authors.

2. "Clever Hans," Wikimedia Foundation, last modified March 12, 2020, https://en.wikipedia.org/wiki/Clever_Hans.

3. Ibid.

4. Edward T. Heyn, "Berlin's Wonderful Horse: He Can Do Almost Everything but Talk," *New York Times*, September 4, 1904, https://timesmachine.nytimes.com/timesmachine/1904/09/04/101396572.pdf.

5. S. Lapuschkin et al., "Analyzing Classifiers: Fisher Vectors and Deep Neural Networks," *IEEE Conference on Computer Vision and Pattern Recognition* (2016): 2912–2920.

6. S. Lapuschkin et al., "Unmasking Clever Hans Predictors and Assessing What Machines Really Learn," *Nature Communications* 10 (March 2019): 1–8.

7. A. Mordvintsev, C. Olah and M. Tyka, "Inceptionism: Going Deeper into Neural Networks," 2015, *Google AI Blog*, https://ai.googleblog.com/2015/06/inceptionism-going-deeper-into-neural.html.

8. See section 23 of the Official Information Act 1982, New Zealand's "freedom of information" legislation: http://www.legislation.govt.nz/act/public/1982/0156/latest/DLM65628.html.

9. Lilian Edwards and Michael Veale, "Slave to the Algorithm? Why a 'Right to an Explanation' Is Probably Not the Remedy You Are Looking For," *Duke Law and Technology Review* 16, no. 1 (2017): 18–84.

10. Jenna Burrell, "How the Machine 'Thinks': Understanding Opacity in Machine Learning Algorithms," *Big Data and Society* 3, no. 1 (2016): 1–12; Edwards and Veale, "Slave to the Algorithm"; Lilian Edwards and Michael Veale, "Enslaving the Algorithm: From a 'Right to an Explanation' to a 'Right To Better Decisions'?" *IEEE Security and Privacy* 16, no. 3 (2018): 46–54; Michael Veale and Lilian Edwards, "Clarity, Surprises, and Further Questions in the Article 29 Working Party Draft Guidance on Automated Decision-Making and Profiling," *Computer Law and Security Review* 34 (2018): 398–404; G. Montavon, S. Bach, A. Binder, W. Samek, and K. R. Müller, "Explaining Nonlinear Classification Decisions with Deep Taylor Decomposition," *Pattern Recognition* 65 (2018): 211; "The IEEE Global Initiative on Ethics of Autonomous and Intelligent Systems," IEEE Standards Association, https://standards.ieee.org/industry-connections/ec/autonomous-systems.html.

11. Edwards and Veale, "Slave to the Algorithm," 64 (emphasis added).

12. "The IEEE Global Initiative on Ethics of Autonomous and Intelligent Systems," IEEE Standards Association, https://standards.ieee.org/industry-connections/ec/autonomous-systems.html, (emphasis added).

13. See, e.g., https://standards.ieee.org/develop/project/7001.html.

14. Regulation (EU) 2016/679 of the European Parliament and of the Council of 27 April 2016 on the protection of natural persons with regard to the processing of personal data and on the free movement of such data, and repealing Directive 95/46/EC (General Data Protection Regulation), OJ L 119, 27.3.2016, p. 1.

15. S. Dutta, "Do computers make better bank managers than humans?" *The Conversation*, October 17, 2017.

16. Brent D. Mittelstadt, Patrick Allo, Mariarosaria Taddeo, Sandra Wachter, and Luciano Floridi, "The Ethics of Algorithms: Mapping the Debate," *Big Data and Society* 16 (2016): 1–21, 7.

17. Luke Muehlhauser, "Transparency in Safety-critical Systems." *Machine Intelligence Research Institute* (blog), August 25, 2013, https://intelligence.org/2013/08/25/transparency-in-safety-critical-systems/.

18. Ronald Dworkin, *Taking Rights Seriously* (London: Duckworth, 1977); Ronald Dworkin, *Law's Empire* (London: Fontana Books, 1986).

19. House of Lords, Select Committee on Artificial Intelligence, "AI in the UK: Ready, Willing, and Able?" April 2018, https://publications.parliament.uk/pa/ld201719/ldselect/ldai/100/100.pdf, 38.

20. Ibid.

21. Ibid., 40 (emphasis added).

22. Ibid., 37.

23. S. Plous, ed. *Understanding Prejudice and Discrimination* (New York: McGraw-Hill, 2003), 2.

24. S. Plous, "The Psychology of Prejudice, Stereotyping, and Discrimination," in *Understanding Prejudice and Discrimination*, ed. S. Plous (New York: McGraw-Hill, 2003), 17.

25. R. McEwen, J. Eldridge, and D. Caruso, "Differential or Deferential to Media? The Effect of Prejudicial Publicity on Judge or Jury," *International Journal of Evidence and Proof* 22, no. 2 (2018): 124–143, 126.

26. Ibid., 136.

27. Ibid., 140.

28. Jeremy Waldron, *The Law* (London: Routledge, 1990).

29. See, e.g., Supreme Court Act, section 101(2) (New South Wales).

30. *Devries v. Australian National Railways Commission* (1993) 177 CLR 472 (High Court of Australia); *Abalos v. Australian Postal Commission* (1990) 171 CLR 167 (High Court of Australia); cf. *Fox v. Percy* (2003) 214 CLR 118 (High Court of Australia).

31. J. C. Pomerol and F. Adam, "Understanding Human Decision Making: A Fundamental Step towards Effective Intelligent Decision Support," in *Intelligent Decision Making: An AI-Based Approach*, ed. G. Phillips-Wren, N. Ichalkaranje, and L. C. Jain (Berlin: Springer, 2008), 24.

32. Ibid.

33. Ibid.

34. M. Piattelli-Palmarini, *La R'eforme du Jugement ou Comment Ne Plus Se Tromper* (Paris: Odile Jacob, 1995); A. Tversky and D. Kahneman, "Judgment Under Uncertainty: Heuristics and Biases," *Science* 185 (1974): 1124–1131.

35. Pomerol and Adam, "Understanding Human Decision Making."

36. J. Pohl, "Cognitive Elements of Human Decision Making," in *Intelligent Decision Making: An AI-Based Approach*, ed. G. Phillips-Wren, N. Ichalkaranje, and L.C. Jain (Berlin: Springer, 2008).

37. Montavon et al., "Explaining Nonlinear Classification Decisions."

38. M. Ribeiro, S. Singh, and C. Guestrin, "'Why Should I Trust You?' Explaining the Predictions of Any Classifier," *Proc. 22nd ACM International Conference on Knowledge Discovery and Data Mining* (2016): 1135–1144.

39. Chaofan Chen et al., "*This* Looks Like *That*: Deep Learning for Interpretable Image Recognition," Preprint, submitted June 27, 2018, https://arxiv.org/pdf/1806.10574.pdf.

40. Ibid., 2 (emphasis added).

41. Alexander Babuta, Marion Oswald, and Christine Rinik, "Machine Learning Algorithms and Police Decision-Making: Legal, Ethical and Regulatory Challenges," *Whitehall Reports* (London: Royal United Services Institute, 2018), 18.

42. Zoe Kleinman, "IBM Launches Tool Aimed at Detecting AI Bias," *BBC*, September 9, 2019.

43. Ibid.

44. John Zerilli, "Explaining Machine Learning Decisions," 2020 (submitted manuscript).

45. E. Langer, A. E. Blank, and B. Chanowitz, "The Mindlessness of Ostensibly Thoughtful Action: The Role of 'Placebic' Information in Interpersonal Interaction," *Journal of Personality and Social Psychology* 36, no. 6 (1978): 635–642.

46. W. M. Oliver and R. Batra, "Standards of Legitimacy in Criminal Negotiations," *Harvard Negotiation Law Review* 20 (2015): 61–120.

Chapter 3

1. A. Tversky and D. Kahneman, "Judgment Under Uncertainty: Heuristics and Biases," *Science* 185 (1974): 1124–1131.

2. Gerd Gigerenzer, Peter M. Todd, and the ABC Research Group, *Simple Heuristics That Make Us Smart* (New York: Oxford University Press, 1999).

3. A. Tversky and D. Kahneman, "Availability: A Heuristic for Judging Frequency and Probability," *Cognitive Psychology* 5, no. 2 (1973): 207–232.

4. G. Loewenstein, *Exotic Preferences: Behavioral Economics and Human Motivation* (New York: Oxford University Press, 2007), 283–284.

5. Endre Begby, "The Epistemology of Prejudice," *Thought: A Journal of Philosophy* 2, no. 1 (2013): 90–99; Sarah-Jane Leslie, "The Original Sin of Cognition: Fear, Prejudice, and Generalization," *Journal of Philosophy* 114, no. 8 (2017): 393–421.

6. S. Lichtenstein, B. Fischoff and L. D. Phillips, "Calibration of Probabilities: The State of the Art to 1980," in *Judgment Under Uncertainty: Heuristics and Biases*, ed. D. Kahneman, P. Slovic, and A. Tversky (Cambridge: Cambridge University Press, 1982).

7. N. Arpaly, *Unprincipled Virtue: An Inquiry into Moral Agency* (New York: Oxford University Press, 2003).

8. Miranda Fricker, *Epistemic Injustice: Power and the Ethics of Knowing* (New York: Oxford University Press, 2007).

9. Begby, "The Epistemology of Prejudice"; Leslie, "The Original Sin of Cognition."

10. J. Pohl, "Cognitive Elements of Human Decision Making," in *Intelligent Decision Making: An AI-Based Approach*, ed. G. Phillips-Wren, N. Ichalkaranje, and L. C. Jain (Berlin: Springer, 2008); A. D. Angie, S. Connelly, E. P. Waples, and V. Kligyte, "The Influence of Discrete Emotions on Judgement and Decision-Making: A Meta-Analytic Review," *Cognition and Emotion* 25, no. 8 (2011): 1393–1422.

11. Cathy O'Neil, *Weapons of Math Destruction: How Big Data Increases Inequality and Threatens Democracy* (New York: Broadway Books, 2016).

12. Toby Walsh, *2062: The World that AI Made* (Melbourne: La Trobe University Press, 2018).

13. Virginia Eubanks, *Automating Inequality: How High-Tech Tools Profile, Police, and Punish the Poor* (New York: St Martin's Press, 2017).

14. J. Larson, S. Mattu, L. Kirchner, and J. Angwin, "How We Analyzed the COMPAS Recidivism Algorithm," *ProPublica*, May 23, 2016, https://www.propublica.org/article/how-we-analyzed-the-compas-recidivism-algorithm.

15. H. Couchman, *Policing by Machine: Predictive Policing and the Threat to Our Rights* (London: Liberty, 2018).

16. Jessica M. Eaglin, "Constructing Recidivism Risk," *Emory Law Journal* 67: 59–122.

17. Lucas D. Introna, "The Enframing of Code," *Theory, Culture and Society* 28, no. 6 (2011): 113–141.

18. T. Petzinger, *Hard Landing: The Epic Contest for Power and Profits that Plunged the Airlines into Chaos* (New York: Random House, 1996).

19. Jamie Bartlett, *The People Vs Tech: How the Internet is Killing Democracy (and How We Save It)* (London: Penguin, 2018); Jamie Susskind, *Future Politics: Living Together in a World Transformed by Tech* (New York: Oxford University Press, 2018).

20. Joseph Turow, *The Daily You: How the New Advertising Industry Is Defining Your Identity and Your Worth* (New Haven: Yale University Press).

21. D. W. Hamilton, "The Evolution of Altruistic Behavior," *American Naturalist* 97, no. 896 (1963): 354–356.

22. H. Tajfel, "Experiments in Intergroup Discrimination," *Scientific American* 223, no. 5 (1970): 96–102.

23. Tim Hughes, "Prediction and Social Investment," in *Social Investment: A New Zealand Policy Experiment*, ed. Jonathan Boston and Derek Gill (Wellington: Bridget Williams Books, 2018), 167.

24. E. Vul and H. Pashler, "Measuring the Crowd Within: Probabilistic Representations within Individuals," *Psychological Science* 19, no. 7 (2008): 645–647.

25. J. Surowiecki, *The Wisdom of Crowds: Why the Many Are Smarter than the Few and How Collective Wisdom Shapes Business, Economies, Societies and Nations* (New York: Doubleday, 2004).

26. Randy Rieland, "Artificial Intelligence Is Now Used to Predict Crime. But Is It Biased?" *Smithsonian*, March 5, 2018.

27. O'Neil, *Weapons of Math Destruction*, 58.

28. American Civil Liberties Union et al., *Predictive Policing Today: A Shared Statement of Civil Rights Concerns*, 2016, https://www.aclu.org/other/statement-concern-about-predictive-policing-aclu-and-16-civil-rights-privacy-racial-justice.

29. Kristian Lum and William Isaac, "To Predict and Serve? Bias in Police-Recorded Data," *Significance* (October 2016): 14–19.

30. Ibid., 16.

31. Andrew D. Selbst, Danah Boyd, Sorelle A. Friedler, Suresh Venkatasubramanian, and Janet Vertesi, "Fairness and Abstraction in Sociotechnical Systems," *Proc. Conference on Fairness, Accountability, and Transparency* (2019): 59–68.

32. Lucas D. Introna, "Maintaining the Reversibility of Foldings: Making the Ethics (Politics) of Information Technology Visible," *Ethics and Information Technology* 9, no. 1 (2006): 11–25.

33. I. Leki and J. Carson, "Completely Different Worlds: EAP and the Writing Experiences of ESL Students in University Courses," *TESOL Quarterly* 31, no.1 (1997): 39–69.

34. D. G. Copeland, R. O. Mason, and J. L. McKenney, "Sabre: The Development of Information-Based Competence and Execution of Information-Based Competition," *IEEE Annals of the History of Computing*, 17, no. 3 (1995): 30–57.

35. Petzinger, *Hard Landing*.

36. Ibid.

37. Benjamin G. Edelman, "Leveraging Market Power Through Tying and Bundling: Does Google Behave Anti-Competitively?" *Harvard Business School* NOM Unit Working Paper, no. 14–112 (2014).

38. D. Mattioli, "On Orbitz, Mac Users Steered to Pricier Hotels," *The Wall Street Journal*, August 23, 2013.

39. G. Neff and P. Nagy, "Talking to Bots: Symbiotic Agency and the Case of Tay," *International Journal of Communication* 10 (2016): 17.

40. P. Mason, "The Racist Hijacking of Microsoft's Chatbot Shows How the Internet Teems with Hate," *The Guardian*, March 29, 2016, https://www.theguardian.com/world/2016/mar/29/microsoft-tay-tweets-antisemitic-racism.

41. Peter Lee, "Learning from Tay's Introduction," *Official Microsoft Blog*, March 25, 2016, https://blogs.microsoft.com/blog/2016/03/25/learning-tays-introduction/#sm.00000gjdpwwcfcus11t6oo6dw79gw.

42. Adam Rose, "Are Face-Detection Cameras Racist?" *Time*, January 22, 2010.

43. D. Harwell, "The Accent Gap," *The Washington Post*, July 19, 2018, https://www.washingtonpost.com/graphics/2018/business/alexa-does-not-understand-your-accent/?utm_term=.3ee60337 6b8e.

44. N. Furl, "Face Recognition Algorithms and the Other-Race Effect: Computational Mechanisms for a Developmental Contact Hypothesis," *Cognitive Science* 26 no. 6 (2002): 797–815.

45. Lucas D. Introna and David Wood, "Picturing Algorithmic Surveillance: The Politics of Facial Recognition Systems," *Surveillance and Society* 2: 177–198.

46. R. Bothwell, J. Brigham and R. Malpass, "Cross-Racial Identification," *Personality and Social Psychology Bulletin* 15 (1985): 19–25.

47. Amanda Levendowski, "How Copyright Law Can Fix Artificial Intelligence's Implicit Bias Problem," *Washington Law Review* 93 (2018): 579–630.

48. Tom Simonite, "Probing the Dark Side of Google's Ad-Targeting System," *MIT Technology Review*, July 6, 2015, https://www.technologyreview.com/s/539021/probing-the-dark-side-of-googles-ad-targeting-system/.

49. Northpointe, *Practitioner's Guide to COMPAS*, 2015, http://www.northpointeinc.com/downloads/compas/Practitioners-Guide-COMPAS-Core-_031915.pdf.

50. Larson et al., "How We Analyzed the COMPAS Recidivism Algorithm."

51. Alexandra Chouldechova, "Fair Prediction with Disparate Impact: A Study of Bias in Recidivism Prediction Instruments," *Big Data* 5, no. 2 (2017): 153–163.

52. Ibid.

53. Sam Corbett-Davies, Emma Pierson, Avi Feller, and Sharad Goel, "A Computer Program Used for Bail and Sentencing Decisions Was Labeled Biased against Blacks. It's Actually Not That Clear," *Washington Post*, October 17, 2016.

Chapter 4

1. H. L. A. Hart, *Punishment and Responsibility: Essays in the Philosophy of Law*, 2nd ed. (New York: Oxford University Press, 2008).

2. Ibid., 211.

3. J. Ladd, "Computers and Moral Responsibility: A Framework for an Ethical Analysis," in *The Information Web: Ethical and Social Implications of Computer Networking*, ed. C. C. Gould (Boulder, CO: Westview Press, 1989); D. Gotterbarn, "Informatics and Professional Responsibility," *Science and Engineering Ethics* 7, no. 2 (2001): 221–230.

4. Deborah G. Johnson, "Computer Systems: Moral Entities but Not moral Agents," *Ethics and Information Technology* 8, no. 4 (November 2006): 195–204.

5. Andrew Eshleman, "Moral Responsibility," in *The Stanford Encyclopedia of Philosophy* ed. Edward N. Zalta, Winter 2016, https://plato.stanford.edu/entries/moral-responsibility/.

6. Maurice Schellekens, "No-Fault Compensation Schemes for Self-Driving Vehicles," *Law, Innovation and Technology* 10, no. 2 (2018): 314–333.

7. Peter Cane, *Responsibility in Law and Morality* (Oxford: Hart Publishing, 2002).

8. Ibid.

9. Karen Yeung, "A Study of the Implications of Advanced Digital Technologies (Including AI Systems) for the Concept of Responsibility Within a Human Rights Framework," Preprint, submitted 2018, https://ssrn.com/abstract=3286027.

10. Carl Mitcham, "Responsibility and Technology: The Expanding Relationship," in *Technology and Responsibility*, ed. Paul T. Durbin (Dordrecht, Netherlands: Springer, 1987).

11. M. Bovens and S. Zouridis, "From Street-Level to System-Level Bureaucracies: How Information and Communication Technology Is Transforming Administrative Discretion and Constitutional Control," *Public Administration Review* 62, no. 2 (2002): 174–184.

12. Hans Jonas, *The Imperative of Responsibility: In Search of an Ethics for the Technological Age* (Chicago: University of Chicago Press, 1984).

13. Andreas Matthias, "The Responsibility Gap: Ascribing Responsibility for the Actions of Learning Automata," *Ethics and Information Technology* 6, no. 3 (September 2004): 175–183.

14. See, e.g., K. Himma, "Artificial Agency, Consciousness, and the Criteria for Moral Agency: What Properties Must an Artificial Agent Have to be a Moral Agent?" *Ethics and Information Technology* 11, no. 1 (2009): 19–29.

15. L. Suchman, "Human/Machine Reconsidered," *Cognitive Studies* 5, no. 1 (1998): 5–13.

16. Gary Marcus, "Innateness, AlphaZero, and Artificial Intelligence," Preprint, submitted January 17, 2018, https://arxiv.org/pdf/1801.05667.pdf.

17. Madeleine C. Elish, "Moral Crumple Zones: Cautionary Tales in Human Robot Interaction," *Engaging Science, Technology and Society* 5 (2019): 40–60.

18. David C. Vladeck, "Machines Without Principals: Liability Rules and Artificial Intelligence," *Washington Law Review* 89 (2014): 117–150.

19. Ibid.

20. Luciano Floridi and J. W. Sanders, "On the Morality of Artificial Agents," *Minds and Machines* 14, no. 3 (August 2004): 349–379.

21. Colin Allen and Wendel Wallach, "Moral Machines: Contradiction in Terms or Abdication of Human Responsibility?" in *Robot Ethics: The Ethical and Social Implications of Robotics*, ed. Patrick Lin, Keith Abney, and George A. Bekey (Cambridge, MA: MIT Press, 2012).

22. Johnson, "Computer Systems"; Deborah G. Johnson and T. M. Power, "Computer Systems and Responsibility: A Normative Look at Technological Complexity," *Ethics and Information Technology* 7, no. 2 (June 2005): 99–107.

23. Peter Kroes and Peter-Paul Verbeek, ed. *The Moral Status of Technical Artefacts* (Dordrecht, Netherlands: Springer, 2014).

24. Peter-Paul Verbeek, "Materializing Morality," *Science, Technology, and Human Values* 31, no. 3 (2006): 361–380.

25. Ugo Pagallo, "Vital, Sophia, and Co.: The Quest for the Legal Personhood of Robots," *Information* 9, no. 9 (2019): 1–11.

Chapter 5

1. Parts of this chapter are reprinted from Springer Nature, Minds and Machines, "Algorithmic Decision-Making and the Control Problem" by John Zerilli, Alistair Knott, James Maclaurin and Colin Gavaghan. Copyright in the authors 2019. As such, it represents the considered views of these authors.

2. C. Villani, *For a Meaningful Artificial Intelligence: Towards a French and European Strategy*, 2018, https://www.aiforhumanity.fr/pdfs/MissionVillani_Report_ENG-VF.pdf.

3. AI Now Institute, *Litigating Algorithms: Challenging Government Use of Algorithmic Decision Systems*, (New York: AI Now Institute, 2018), https://ainowinstitute.org/litigatingalgorithms.pdf.

4. Ibid.

5. Virginia Eubanks, *Automating Inequality: How High-Tech Tools Profile, Police, and Punish the Poor* (New York: St Martin's Press, 2017).

6. Northpointe, *Practitioner's Guide to COMPAS,* 2015, http://www.northpointeinc.com/downloads/compas/Practitioners-Guide-COMPAS-Core-_031915.pdf.

7. *Wisconsin v. Loomis* 881 N.W.2d 749, 123 (Wis. 2016).

8. Ibid., 100.

9. Raja Parasuraman and Dietrich H. Manzey, "Complacency and Bias in Human Use of Automation: An Attentional Integration," *Human Factors* 52, no. 3 (June 2010): 381–410.

10. C. D. Wickens and C. Kessel, "The Effect of Participatory Mode and Task Workload on the Detection of Dynamic System Failures," *IEEE Trans. Syst., Man, Cybern.* 9, no. 1 (January 1979): 24–31; E. L. Wiener and R. E. Curry, "Flight-Deck Automation: Promises and Problems," *Ergonomics* 23, no. 10 (1980): 995–1011.

11. Lisanne Bainbridge, "Ironies of Automation," *Automatica* 19, no. 6 (1983): 775–779, 775.

12. Ibid., 776 (emphasis added).

13. Ibid.

14. Gordon Baxter, John Rooksby, Yuanzhi Wang, and Ali Khajeh-Hosseini, "The Ironies of Automation … Still Going Strong at 30?" *Proc. ECCE Conference Edinburgh* (2012): 65–71, 68.

15. David Cebon, "Responses to Autonomous Vehicles," *Ingenia* 62 (March 2015): 10.

16. Bainbridge, "Ironies," 776.

17. Neville A. Stanton, "Distributed Situation Awareness," *Theoretical Issues in Ergonomics Science* 17, no. 1 (2016): 1–7.

18. Neville A. Stanton, "Responses to Autonomous Vehicles," *Ingenia* 62 (March 2015): 9; Mitchell Cunningham and Michael Regan, "Automated Vehicles May Encourage a New Breed of Distracted Drivers," *The Conversation*, September 25, 2018; Victoria A. Banks, Alexander Erikssona, Jim O'Donoghue, and Neville A. Stanton, "Is Partially Automated Driving a Bad Idea? Observations from an On-Road Study," *Applied Ergonomics* 68 (2018): 138–145; Victoria A. Banks, Katherine L. Plant, and Neville A. Stanton, "Driver Error or Designer Error: Using the Perceptual Cycle Model to Explore the Circumstances Surrounding the Fatal Tesla Crash on 7th May 2016," *Safety Science* 108 (2018): 278–285.

19. Bainbridge, "Ironies," 775.

20. Ibid., 776.

21. Wiener and Curry, "Flight-Deck Automation."

22. See, e.g., Linda J. Skitka, Kathleen Mosier and Mark D. Burdick, "Accountability and Automation Bias," *International Journal of Human-Computer Studies* 52 (2000): 701–717; Parasuraman and Manzey, "Complacency and Bias"; Kayvan Pazouki, Neil Forbes, Rosemary A. Norman, and Michael D. Woodward, "Investigation on the Impact of Human-Automation Interaction in Maritime Operations," *Ocean Engineering* 153 (2018): 297–304.

23. Pazouki et al., "Investigation," 299.

24. Ibid.

25. Parasuraman and Manzey, "Complacency and Bias," 406.

26. Stanton, "Responses to Autonomous Vehicles."

27. Banks et al., "Driver Error or Designer Error," 283.

28. Ibid.

29. N. Bagheri and G. A. Jamieson, "Considering Subjective Trust and Monitoring Behavior in Assessing Automation-Induced 'Complacency'," in *Human Performance, Situation Awareness, and Automation: Current Research and Trends*, ed. D. A. Vicenzi, M. Mouloua, and O. A. Hancock (Mahwah, NJ: Erlbaum, 2004).

30. Banks et al., "Driver Error or Designer Error," 283.

31. J. Pohl, "Cognitive Elements of Human Decision Making," in *Intelligent Decision Making: An AI-Based Approach*, ed. G. Phillips-Wren, N. Ichalkaranje and L. C. Jain (Berlin: Springer, 2008).

32. Banks et al., "Is Partially Automated Driving a Bad Idea?"; Banks et al., "Driver Error or Designer Error."

33. Guy H. Walker, Neville A. Stanton, and Paul M. Salmon, *Human Factors in Automotive Engineering and Technology* (Surrey: Ashgate, 2015).

34. Mark Bridge, "AI Can Identify Alzheimer's Disease a Decade before Symptoms Appear," *The Times*, September 20, 2017.

35. Nikolaos Aletras, Dimitrios Tsarapatsanis, Daniel Preotiuc-Pietro, and Vasileios Lampos, "Predicting Judicial Decisions of the European Court of Human Rights: A Natural Language Processing Perspective," *PeerJ Computer Science* 2, no. 93 (October 2016): 1–19.

36. Erik Brynjolfsson and Andrew McAfee, *Machine Platform Crowd: Harnessing Our Digital Future* (New York: Norton, 2017).

37. Skitka et al., "Accountability and Automation Bias," 701.

38. Parasuraman and Manzey, "Complacency and Bias," 392.

Chapter 6

1. Samuel D. Warren and Louis D. Brandeis, "The Right to Privacy," *Harvard Law Review* 4, no. 5 (December 1890): 193–220.

2. Universal Declaration of Human Rights (1948).

3. Julie C. Inness, *Privacy, Intimacy and Isolation* (New York: Oxford University Press, 1992).

4. Ibid.

5. Daniel J. Solove, "A Taxonomy of Privacy," *University of Pennsylvania Law Review* 154, no. 3 (January 2006): 477–560.

6. David Banisar and Simon Davies, *Privacy and Human Rights: An International Survey of Privacy Law and Developments* (Global Internet Liberty Campaign, 2000), http://gilc.org/privacy/survey/intro.html.

7. Lilian Edwards and Michael Veale, "Slave to the Algorithm? Why a 'Right to an Explanation' Is Probably Not the Remedy You Are Looking For," *Duke Law and Technology Review* 16, no. 1 (2017): 18–84, 32.

8. Privacy International, *Privacy and Freedom of Expression in the Age of Artificial Intelligence* (London: 2018).

9. David Locke and Karen Lewis, "The Anatomy of an IoT Solution: Oli, Data and the Humble Washing Machine," October 17, 2017, https://www.ibm.com/blogs/internet-of-things/washing-iot-solution/.

10. C. Epp, M. Lippold, and R.L. Mandryk, "Identifying Emotional States Using Keystroke Dynamics," *Proc. SIGHI Conference on Human Factors in Computing Systems* (2011): 715–724.

11. Yulin Wang and Michal Kosinski, "Deep Neural Networks Are More Accurate than Humans at Detecting Sexual Orientation from Facial Images," *Journal of Personality and Social Psychology* 114, no. 2 (2018): 246–257.

12. Blaise Agüera y Arcas, Alexander Todorov and Margaret Mitchell, "Do Algorithms Reveal Sexual Orientation or Just Expose Our Stereotypes?" *Medium*, January 11, 2018, https://medium.com/@blaisea/do-algorithms-reveal-sexual-orientation-or-just-expose-our-stereotypes-d998fafdf477.

13. John Leuner, "A Replication Study: Machine Learning Models Are Capable of Predicting Sexual Orientation from Facial Images," February 2019, Preprint, https://arxiv.org/pdf/1902.10739.pdf.

14. Paul Ohm, "Broken Promises of Privacy: Responding to the Surprising Failure of Anonymization," *UCLA Law Review* (2010) 57: 1701–1777.

15. J.P. Achara, G. Acs and C. Castelluccia, "On the Unicity of Smartphone Applications" *Proc. 14th ACM Workshop on Privacy in the Electronic Society* (2015): 27–36.

16. US Department of Housing and Urban Development, "HUD Charges Facebook with Housing Discrimination over Company's Targeted Advertising Practices," HUD press release no. 19–035, March 28, 2019.

17. Katie Benner, Glenn Thrush, and Mike Isaac, "Facebook Engages in Housing Discrimination with Its Ad Practices, US Says," *New York Times*, March 28, 2019.

18. Ibid.

19. Virginia Eubanks, *Automating Inequality: How High-Tech Tools Profile, Police, and Punish the Poor* (New York: St Martin's Press, 2017).

20. Salesforce and Deloitte, *Consumer Experience in the Retail Renaissance*, 2018, https://c1.sfdcstatic.com/content/dam/web/en_us/www/documents/e-books/learn/consumer-experience-in-the-retail-renaissance.pdf.

21. Elizabeth Denham, *Investigation into the Use of Data Analytics in Political Campaigns: A Report to Parliament* (London: Information Commissioner's Office, 2018).

22. Ibid., 4. See also Information Commissioner's Office, *Democracy Disrupted? Personal Information and Political Influence* (London: Information Commissioner's Office, 2018).

23. Information Commissioner's Office, *Democracy Disrupted.*

24. Mary Madden and Lee Raine, *Americans' Attitudes about Privacy, Security and Surveillance* (Washington: Pew Research Center, 2015), https://www.pewinternet.org/2015/05/20/americans-attitudes-about-privacy-security-and-surveillance/.

25. Ann Couvakian, "7 Foundational Principles," *Privacy by Design* (Ontario: Information and Privacy Commissioner, 2009), https://www.ipc.on.ca/wp-content/uploads/Resources/7foundationalprinciples.pdf.

26. Sandra Wachter and Brent D. Mittelstadt, "A Right to Reasonable Inferences: Re-thinking Data Protection Law in the Age of Big Data and AI" *Columbia Business Law Review* (forthcoming).

27. Article 29 Data Protection Working Party, "Opinion 4/2007 on the Concept of Personal Data," 01248/07/EN (June 20, 2007), https://ec.europa.eu/justice/article-29/documentation/opinion-recommendation/files/2007/wp136_en.pdf.

Chapter 7

1. Evgeny Morozov, "The Real Privacy Problem," *MIT Technology Review*, October 22, 2013, http://www.technologyreview.com/featuredstory/520426/the-real-privacy-problem/.

2. Yuval Noah Harari, "Liberty," in *21 Lessons for the 21st Century* (London: Harvill Secker, 2018).

3. For example, Suzy Killmister, *Taking the Measure of Autonomy: A Four-Dimensional Theory of Self-Governance* (London: Routledge, 2017) and Quentin Skinner, "The Genealogy of Liberty," Public Lecture, UC Berkley, September 15, 2008, video, 1:17:03, https://www.youtube.com/watch?v=ECiVz_zRj7A.

4. Joseph Raz, *The Morality of Freedom* (Oxford: Oxford University Press, 1986), 373.

5. Barry Schwartz, *The Paradox of Choice: Why Less Is More* (New York: Harper Collins, 2004).

6. Philip Pettit, *Republicanism: A Theory of Freedom and Government* (Oxford: Oxford University Press, 2001); Philip Pettit, "The Instability of Freedom as Non-Interference: The Case of Isaiah Berlin," *Ethics* 121, no. 4 (2011): 693–716; Philip Pettit, *Just Freedom: A Moral Compass for a Complex World* (New York: Norton, 2014).

7. John Danaher, "Moral Freedom and Moral Enhancement: A Critique of the 'Little Alex' Problem," in *Royal Institute of Philosophy Supplement on Moral Enhancement*, ed. Michael Hauskeller and Lewis Coyne (Cambridge: Cambridge University Press, 2018).

8. Brett Frischmann and Evan Selinger, *Re-Engineering Humanity* (Cambridge: Cambridge University Press, 2018).

9. C. T. Nguyen, "Echo Chambers and Epistemic Bubbles" *Episteme* (forthcoming), https://doi.org/10.1017/epi.2018.32.

10. David Sumpter, *Outnumbered: From Facebook and Google to Fake News and Filter-Bubbles: The Algorithms that Control Our Lives* (London: Bloomsbury Sigma, 2018).

11. Jamie Susskind, *Future Politics: Living Together in a World Transformed by Tech* (New York: Oxford University Press, 2018).

12. Ibid., 347.

13. Susskind, *Future Politics.*

14. J. Matthew Hoye and Jeffrey Monaghan, "Surveillance, Freedom and the Republic," *European Journal of Political Theory,* 17, no. 3 (2018): 343–363.

15. For a critical meta-analysis of this phenomenon, see B. Scheibehenne, R. Greifeneder, and P. M. Todd, "Can There Ever Be Too Many Options? A Meta-Analytic Review of Choice Overload," *Journal of Consumer Research,* 37 (2010): 409–425.

16. Nick Bostrom and Toby Ord, "The Reversal Test: Eliminating Status Quo Bias in Applied Ethics," *Ethics* 116 (2006): 656–679.

17. Shoshana Zuboff, *The Age of Surveillance Capitalism* (London: Profile Books, 2019).

18. Rogier Creemers, "China's Social Credit System: An Evolving Practice of Control," May 9, 2018, https://ssrn.com/abstract=3175792.

19. Richard Thaler and Cass Sunstein, *Nudge: Improving Decisions about Health, Wealth and Happiness* (London: Penguin, 2009).

20. Daniel Kahneman, *Thinking, Fast and Slow* (New York: Farrar, Straus and Giroux, 2011).

21. Cass Sunstein, *The Ethics of Influence* (Cambridge: Cambridge University Press, 2016).

22. Karen Yeung, "'Hypernudge': Big Data as a Mode of Regulation By Design," *Information, Communication and Society* 20, no. 1 (2017): 118–136; Marjolein Lanzing, "'Strongly Recommended': Revisiting Decisional Privacy to Judge Hypernudging in Self-Tracking Technologies," *Philosophy and Technology* (2018), https://doi.org/10.1007/s13347-018-0316-4.

23. Tom O'Shea, "Disability and Domination," *Journal of Applied Philosophy* 35, no. 1 (2018): 133–148.

24. Janet Vertesi, "Internet Privacy and What Happens When You Try to Opt Out," *Time,* May 1, 2014.

25. Angèle Christin, "Counting Clicks. Quantification and Variation in Web Journalism in the United States and France," *American Journal of Sociology* 123, no. 5

(2018): 1382–1415; Angèle Christin, "Algorithms in Practice: Comparing Web Journalism and Criminal Justice," *Big Data and Society* 4, no. 2 (2017): 1–14.

26. Alexandre Bovet and Hernán A. Makse, "Influence of Fake News in Twitter during the 2016 US Presidential Election," *Nature Communications* 10 (2019), https://www.nature.com/articles/s41467-018-07761-2.

27. Andrew Guess, Brendan Nyhan, Benjamin Lyons, and Jason Reifler, *Avoiding the Echo Chamber about Echo Chambers*, Knight Foundation White Paper, 2018, https://kf-site-production.s3.amazonaws.com/media_elements/files/000/000/133/original/Topos_KF_White-Paper_Nyhan_V1.pdf; Andrew Guess, Jonathan Nagler, and Joshua Tucker, "Less Than You Think: Prevalence and Predictors of Fake News Dissemination on Facebook," *Science Advances* 5, no. 1 (Jan 2019).

28. Michele Loi and Paul Olivier DeHaye, "If Data Is the New Oil, When Is the Extraction of Value from Data Unjust?" *Philosophy and Public Issues* (New Series) 7, no. 2 (2017): 137–178.

29. Mark Zuckerberg, "A Blueprint for Content Governance and Enforcement," November 15, 2018, https://m.facebook.com/notes/mark-zuckerberg/a-blueprint-for-content-governance-and-enforcement/10156443129621634/.

30. Frischmann and Selinger, *Re-Engineering Humanity*, 270–271.

31. Susskind, *Future Politics*.

32. For an excellent historical overview of how autonomy became central to European Enlightenment thinking, see J. B. Schneewind, *The Invention of Autonomy* (Cambridge: Cambridge University Press, 1998).

Chapter 8

1. Adam Smith, "Of the Division of Labour" in *On the Wealth of Nations* (London: Strahan and Cadell, 1776).

2. Thomas Malone, *Superminds: The Surprising Power of People and Computers Thinking Together* (London: Oneworld Publications, 2018); Joseph Henrich, *The Secret of Our Success* (Princeton, NJ: Princeton University Press 2015).

3. Fabienne Peter, "Political Legitimacy," in *The Stanford Encyclopedia of Philosophy*, ed. Edward N. Zalta, Spring 2017, https://plato.stanford.edu/entries/legitimacy/; John Danaher, "The Threat of Algocracy: Reality, Resistance and Accommodation," *Philosophy and Technology* 29, no. 3 (2016): 245–268.

4. Paul Tucker, *Unelected Power: The Quest for Legitimacy in Central Banking and the Regulatory State* (Princeton, NJ: Princeton University Press, 2018).

5. This story is pieced together from several different sources, specifically, "Gardai Renewed Contract for Speed Vans that 'Should Be Consigned to the Dustbin,'" *The Journal*, October 16, 2016, https://www.thejournal.ie/gosafe-speed-camera-van-2-2715

594-Apr2016/; "Donegal Judge Dismisses Go-Safe Van Speeding Cases," *Donegal News*, December 3, 2014, https://donegalnews.com/2014/12/donegal-judge-dismisses-go-safe-van-speeding-cases/; Gordon Deegan, "Judge Asks Are Men in Speed Camera Vans Reading Comic Books," *The Irish Times*, March 22, 2014, https://www.irishtimes.com/news/crime-and-law/courts/judge-asks-are-men-in-speed-camera-vans-reading-comic-books-1.1734313; Wayne O'Connor Judge, "Go Safe Speed Camera Vans Bring Law into Disrepute," *The Irish Independent*, December 4, 2014, https://www.independent.ie/irish-news/courts/judge-go-safe-speed-camera-vans-bring-law-into-disrepute-30797457.html; Edwin McGreal, "New Loophole Uncovered in Go Safe Prosecutions," *Mayo News*, April 28, 2015, http://www.mayonews.ie/component/content/article?id=21842:new-loophole-uncovered-in-go-safe-prosecutions. It is also based on two Irish legal judgments: *Director of Public Prosecutions v. Brown* [2018] IEHC 471; and *Director of Public Prosecutions v. Gilvarry* [2014] IEHC 345.

6. We should note that some people dispute whether speeding is a major cause of road death or at least suggest that its contribution is overstated. A good overview of this, with a specific focus on the attitude toward automatic speed cameras in Ireland and the UK can be found in Anthony Behan, "The Politics of Technology: An Assessment of the Barriers to Law Enforcement Automation in Ireland," (master's thesis, National University of Ireland, Cork, 2016), https://www.academia.edu/32662269/The_Politics_of_Technology_An_Assessment_of_the_Barriers_to_Law_Enforcement_Automation_in_Ireland.

7. "Gardai Renewed Contract."

8. "'Motorists Were Wrongly Fined': Speed Camera Whistleblower," *The Journal*, April 1, 2014, https://www.thejournal.ie/wrongly-fined-1393534-Apr2014/.

9. "GoSafe speed-camera system an 'abject failure': Judge Devins," *Mayo News* March 20, 2012, http://www.mayonews.ie/component/content/article?id=14888:gosafe-speed-camera-system-an-abject-failure-judge-devins.

10. Some of these had to do with legal uncertainty about how to treat the evidence. Some of this uncertainty was addressed by the Irish courts in *Director of Public Prosecutions v. Brown* [2018] IEHC 471 and *Director of Public Prosecutions v. Gilvarry* [2014] IEHC 345.

11. "Gardai Renewed Contract."

12. Cary Coglianese and David Lehr, "Regulating by Robot: Administrative Decision Making in the Machine-Learning Era," *The Georgetown Law Journal* 105 (2017): 1147–1223; Marion Oswald, "Algorithm-Assisted Decision-Making in the Public Sector: Framing the Issues Using Administrative Law Rules Governing Discretionary Power," *Philosophical Transactions of the Royal Society A* 376 (2018): 1–20.

13. Coglianese and Lehr, "Regulating by Robot," 1170.

14. Michael Wellman and Uday Rajan, "Ethical Issues for Autonomous Agents," *Minds and Machines* 27, no. 4 (2017): 609–624.

15. Coglianese and Lehr, "Regulating by Robot," 1178.

16. Ibid., 1182–1184.

17. Oswald, "Algorithm-Assisted Decision-Making in the Public Sector," 14.

18. Coglianese and Lehr, "Regulating by Robot," 1185.

19. Virginia Eubanks, *Automating Inequality* (New York: St Martin's Press, 2017); Dan Hurley, "Can an Algorithm Tell When Kids Are in Danger?" *New York Times*, January 2, 2018, https://www.nytimes.com/2018/01/02/magazine/can-an-algorithm-tell-when-kids-are-in-danger.html; "The Allegheny Family Screening Tool," Allegheny County, https://www.alleghenycountyanalytics.us/wp-content/uploads/2017/07/AFST-Frequently-Asked-Questions.pdf.

20. Rhema Vaithianathan, Bénédicte Rouland, and Emily Putnam-Hornstein, "Injury and Mortality Among Children Identified as at High Risk of Maltreatment," *Pediatrics* 141 no. 2 (February 2018): e20172882.

21. Teuila Fuatai, " 'Unprecedented Breaches of Human Rights': The Oranga Tamariki Inquiry Releases Its Findings," *Spinoff*, February 4, 2020, https://thespinoff.co.nz/atea/numa/04-02-2020/unprecedented-breaches-of-human-rights-the-oranga-tamariki-inquiry-releases-its-findings/.

22. Virginia Eubanks, *Automating Inequality*, 138.

23. Stacey Kirk, "Children 'Not Lab-Rats'—Anne Tolley Intervenes in Child Abuse Experiment," Stuff, July 30, 2015, https://www.stuff.co.nz/national/health/70647353/children-not-lab-rats---anne-tolley-intervenes-in-child-abuse-experiment.

24. "Allegheny Family."

25. Tim Dare and Eileen Gambrill, "Ethical Analysis: Predictive Risk Models at Call Screening for Allegheny County," *Ethical Analysis: DHS Response*, (Allegheny County, PA: Allegheny County Department of Human Services, 2017), https://www.alleghenycounty.us/WorkArea/linkit.aspx?LinkIdentifier=id&ItemID=6442457401.

26. "Frequently Asked Questions" Allegheny County, last modified July 20, 2017, https://www.alleghenycountyanalytics.us/wp-content/uploads/2017/07/AFST-Frequently-Asked-Questions.pdf/.

27. Virginia Eubanks, "The Allegheny Algorithm," *Automating Inequality*.

28. Virginia Eubanks, "A Response to the Allegheny County DHS," *Virginia Eubanks* (blog), February 6, 2018, https://virginia-eubanks.com/2018/02/16/a-response-to-allegheny-county-dhs/.

29. Evgeny Morozov, *To Save Everything Click Here* (New York: Public Affairs, 2013).

30. Dare and Gambrill, "Ethical Analysis," 5–7.

31. Joseph Tainter, *The Collapse of Complex Societies* (Cambridge: Cambridge University Press, 1988).

32. Guy Middleton, *Understanding Collapse: Ancient History and Modern Myths* (Cambridge: Cambridge University Press, 2011).

33. Miles Brundage, "Scaling Up Humanity: The Case for Conditional Optimism about AI," in *Should We Fear the Future of Artificial Intelligence?* European Parliamentary Research Service (2018). For a similar, though slightly more pessimistic, argument, see Phil Torres, "Superintelligence and the Future of Governance: On Prioritizing the Control Problem at the End of History," in *Artificial Intelligence Safety and Security*, ed. Roman V. Yampolskiy (Boca Raton, FL: Chapman and Hall/CRC Press, 2017).

Chapter 9

1. For a good example, see the section headed "Seven Deadly Trends" from Tom Ford's *Rise of the Robots: Technology and the Threat of a Jobless Future* (New York: Basic Books, 2015), 35–61.

2. *The Economic Report of the President* (Washington, D.C.: Chair of the Council of Economic Advisers, 2013), table B-47.

3. Thomas Piketty, *Capital in the Twenty-First Century* (Cambridge, MA: Harvard University Press, 2014).

4. J. Mokyr, *The Enlightened Economy: Britain and the Industrial Revolution 1700—1850* (London: Penguin, 2009).

5. E. Brynjolfsson and A. McAfee, *The Second Machine Age: Work, Progress, and Prosperity in a Time of Brilliant Technologies* (New York: Norton, 2014).

6. Helmut Küchenhoff, "The Diminution of Physical Stature of the British Male Population in the 18th-Century," *Cliometrica* 6, no. 1 (2012): 45–62.

7. R. C. Allen, "Engels' Pause: Technical Change, Capital Accumulation, and Inequality in the British Industrial Revolution," *Explorations in Economic History* 46, no. 4 (2009): 418–435.

8. John M. Keynes, *Essays in Persuasion* (London: Macmillan, 1931), 358–374.

9. Jeremy Rifkin, *The End of Work: The Decline of the Global Labor Force and the Dawn of the Post-Market Era* (New York: Putnam Publishing Group, 1995).

10. Carl B. Frey and Michael A. Osborne, "The Future of Employment: How Susceptible Are Jobs to Computerisation?" (working paper, Oxford Martin School, University

of Oxford, 2013), https://www.oxfordmartin.ox.ac.uk/downloads/academic/The_Future_of_Employment.pdf.

11. M. Artnz, T. Gregory, and U. Ziehran, "The Risk of Automation for Jobs in OECD Countries," OECD Social, Employment, and Migration Working Papers, no. 189 (2016), https://www.keepeek.com//Digital-Asset-Management/oecd/socialissues-migration-health/the-risk-of-automation-for-jobs-in-oecd-countries_5jlz9h56dvq7en#page1.

12. For a detailed discussion see *The Impact of Artificial Intelligence on Work: An Evidence Review Prepared for the Royal Society and the British Academy* (London: Frontier Economics, 2018), section 3.2.2, https://royalsociety.org/-/media/policy/projects/ai-and-work/frontier-review-the-impact-of-AI-on-work.pdf.

13. "Why Women Still Earn Much Less than Men," in *Seriously Curious: The Economist Explains the Facts and Figures that Turn Your World Upside Down*, ed. Tom Standage (London: Profile Books, 2018), 115.

14. Katharine McKinnon, "Yes, AI May Take Some Jobs—But It Could Also Mean More Men Doing Care Work," *The Conversation*, September 13, 2018.

15. "Why Women Still Earn Much Less than Men," 115.

16. Fabrizio Carmignani, "Women Are Less Likely to Be Replaced by Robots and Might Even Benefit from Automation," *The Conversation*, May 17, 2018.

17. McKinnon, "Yes, AI May Take Some Jobs."

18. Ibid.

19. M. Goos and A. Manning, "Lousy and Lovely Jobs: The Rising Polarization of Work in Britain," *The Review of Economics and Statistics* 89, no. 1 (2004): 118–133.

20. Daron Acemoglu and David Autor, "Skills, Tasks and Technologies: Implications for Employment and Earnings," in *Handbook of Labor Economics*, vol. 4B, ed. Orley Ashenfelter and David Card (Amsterdam: Elsevier, 2011).

21. Daron Acemoglu and Pascual Restrepo, "Robots and Jobs: Evidence from US Labor Markets," National Bureau of Economic Research Working Paper, no. 23285, Cambridge, MA, March 2017.

22. S. Kessler, *Gigged: The Gig Economy, the End of the Job and the Future of Work* (London: Random House, 2018).

23. A. Rosenblat and L. Stark, "Algorithmic Labor and Information Asymmetries: A Case Study of Uber's Drivers," *International Journal of Communication* 10 (2016): 3758–3784.

24. Jack Shenker, "Strike 2.0: The Digital Uprising in the Workplace," *The Guardian Review*, August 31, 2019, 9.

25. Ibid.

26. Adam Greenfield, *Radical Technologies: The Design of Everyday Life* (London: Verso, 2017), 199.

27. For a good discussion of the risks of algorithm use in the gig economy see Jeremias Prassl's *Humans As a Service: The Promise and Perils of Work in the Gig Economy* (New York: Oxford University Press, 2018).

28. Shenker, "Strike 2.0," 10.

29. Bertrand Russell, *In Praise of Idleness and Other Essays* (London: Allen and Unwin, 1932), 23.

30. Ibid., 16.

31. This argument is nicely set out in Kate Raworth's much-cited *Doughnut Economics: Seven Ways to Think Like a 21st-Century Economist* (White River Junction, VT: Chelsea Green Publishing, 2017).

32. "OECD Average Annual Hours Actually Worked per Worker," Statistics, Organization for Economic Co-operation and Development, last modified March 16, 2020, https://stats.oecd.org/Index.aspx?DataSetCode=ANHRS.

33. Russell, *In Praise of Idleness.*

34. Bertrand Russell, *The Conquest of Happiness* (New York: Liveright, 1930).

35. Ibid.

Chapter 10

1. "Elon Musk: Artificial Intelligence Is Our Biggest Existential Threat," *The Guardian*, October 27, 2014, https://www.theguardian.com/technology/2014/oct/27/elon-musk-artificial-intelligence-ai-biggest-existential-threat.

2. Amanda Macias, "Facebook CEO Mark Zuckerberg Calls for More Regulation of Online Content," *CNBC*, Fenruary 15, 2020, https://www.cnbc.com/2020/02/15/facebook-ceo-zuckerberg-calls-for-more-government-regulation-online-content.html.

3. Cathy Cobey, "AI Regulation: It's Time for Training Wheels," *Forbes*, March 2019, https://www.forbes.com/sites/insights-intelai/2019/03/27/ai-regulation-its-time-for-training-wheels/#51c8af952f26.

4. James Arvanitakis, "What are Tech Companies Doing about Ethical Use of Data? Not Much," *The Conversation*, November 28, 2018, https://theconversation.com/what-are-tech-companies-doing-about-ethical-use-of-data-not-much-104845.

5. Jacob Turner, *Robot Rules: Regulating Artificial Intelligence* (London: Palgrave Macmillan, 2019), 210.

6. Turner, *Robot Rules*, 212–213.

7. Google, "Perspectives on Issues in AI Governance," 2019, https://ai.google/static/documents/perspectives-on-issues-in-ai-governance.pdf.

8. Ibid.

9. Roger Brownsword and Morag Goodwin, *Law and the Technologies of the Twenty-First Century* (Cambridge: Cambridge University Press, 2016).

10. "Autonomous Weapons: An Open Letter from AI and Robotics Researchers," Future of Life Institute, July 28, 2015, https://futureoflife.org/open-letter-autonomous-weapons.

11. Anita Chabrita, "California Could Soon Ban Facial Recognition Technology on Police Body Scanners," *Los Angeles Times*, September 12, 2019.

12. Shirin Ghaffary, "San Franciscos's Facial Recognition Technology Ban, Explained," *Vox,* May 14, 2019, https://www.vox.com/recode/2019/5/14/18623897/san-francisco-facial-recognition-ban-explained.

13. Renee Diresta, "A New Law Makes Bots Identify Themselves: That's the Problem," *Wired*, July 24, 2019, https://www.wired.com/story/law-makes-bots-identify-themselves.

14. "OECD Principles on AI," *Going Digital*, June, 2019, https://www.oecd.org/going-digital/ai/principles/.

15. "Beijng AI Principles," BAAI (blog), May 29, 2019, https://www.baai.ac.cn/blog/beijing-ai-principles.

16. House of Lords Select Committee on Artificial Intelligence, "AI in the UK: Ready, Willing and Able?" April 2018, https://publications.parliament.uk/pa/ld201719/ldselect/ldai/100/100.pdf.

17. The Japanese Society for Artificial Intelligence, *Ethical Guidelines*, May 2017, http://ai-elsi.org/wp-content/uploads/2017/05/JSAI-Ethical-Guidelines-1.pdf.

18. "Asilomar AI Principles," Future of Life Institute, 2017, https://futureoflife.org/ai-principles.

19. "Microsoft AI Principles," Microsoft Corporation, November 2018, https://www.microsoft.com/en-us/ai/our-approach-to-ai; Sundar Pichai, "AI at Google: Our Principles," *Google AI* (blog), June 7, 2018, https://www.blog.google/technology/ai/ai-principles/.

20. House of Lords Select Committee, 13.

21. Gregory N. Mandel, "Emerging Technology Governance," in *Innovative Governance Models for Emerging Technologies*, ed. Gary Marchant, Kenneth Abbot, and Braden Allenby (Cheltenham: Edward Elgar Publishing, 2013), 62.

22. Cass R. Sunstein, *Laws of Fear: Beyond the Precautionary Principle* (Cambridge: Cambridge University Press, 2005), 58.

23. European Parliament Resolution of 16 February 2017 with Recommendations to the Commission on Civil Law Rules on Robotics (2015/2103(INL)), http://www.europarl.europa.eu/doceo/document/TA-8-2017-0051_EN.html.

24. Nick Bostrom, "Existential Risks: Analyzing Human Extinction Scenarios and Related Hazards," *Journal of Evolution and Technology* 9 (2002): 3.

25. Ibid, 25.

26. Ben Rooney, "Women and Children First: Technology and Moral Panic," *Wall Street Journal*, June 11, 2011.

27. Eric E. Johnson, "The Black Hole Case: The Injunction against the End of the World," *Tennessee Law Review* 76, no. 4 (2009): 819–908.

28. Karen Hao, "Here Are 10 Ways AI Could Help Fight Climate Change," *MIT Technology Review*, June 20, 2019, https://www.technologyreview.com/s/613838/ai-climate-change-machine-learning/.

29. Scherer, "Regulating Artificial Intelligence Systems."

30. Andrea Coravos, Irene Chen, Ankit Gordhandas, and Ariel D. Stern, "We Should Treat Algorithms Like Prescription Drugs," *Quartz*, February 15, 2019; Olaf J. Groth, Mark J. Nitzberg, and Stuart J. Russell, "AI Algorithms Need FDA-Style Drug Trials," *Wired*, August 15, 2019; Andrew Tutt, "An FDA for Algorithms," *Administrative Law Review* 69, no. 1 (2017): 83.

31. Groth et al., "AI Algorithms Need FDA-Style Drug Trials."

32. Coravos et al., "We Should Treat Algorithms Like Prescription Drugs."

33. Ibid.

34. Colin Gavaghan, Alistair Knott, James Maclaurin, John Zerilli and Joy Liddicoat, *Government Use of Artificial Intelligence in New Zealand* (Wellington: New Zealand Law Foundation, 2019).

35. Tutt, "An FDA for Algorithms."

36. Michael Taylor, "Self-Driving Mercedes-Benzes Will Prioritize Occupant Safety over Pedestrians," *Car and Driver*, October 7, 2016, https://www.caranddriver.com/news/a15344706/self-driving-mercedes-will-prioritize-occupant-safety-over-pedestrians/.

37. Ibid.

38. Jean-François Bonnefon, Azim Shariff, and Iyad Rahwan, "The Social Dilemma of Autonomous Vehicles," *Science* 352, no. 6293 (June 2016): 1573–1576.

39. Ibid, 1575.

40. Edmond Awad, Sohan Dsouza, Richard Kim, Jonathan Schulz, Joseph Henrich, Azim Shariff, Jean-François Bonnefon, and Iyad Rahwan, "The Moral Machine Experiment," *Nature* 563 (2018): 59–64.

41. Federal Ministry of Transport and Digital Infrastructure, "Automated and Connected Driving," *Ethics Commission Report*, June 20, 2017.

Epilogue

1. Javier Espinoza, "Coronavirus Prompts Delays and Overhaul of EU Digital Strategy," *Financial Times*, March 22, 2020.

2. Pascale Davies, "Spain Plans Universal Basic Income to Fix Coronavirus Economic Crisis," *Forbes*, April 6, 2020, https://www.forbes.com/sites/pascaledavies/2020/04/06/spain-aims-to-roll-out-universal-basic-income-to-fix-coronavirus-economic-crisis/#68d9f7474b35.

Index